THE GENERATION OF
HIGH MAGNETIC FIELDS

THE INTERNATIONAL CRYOGENICS MONOGRAPH SERIES

General Editors

Dr. K. Mendelssohn, F. R. S.
The Clarendon Laboratory
Oxford, England

Dr. K. D. Timmerhaus
University of Colorado
Boulder, Colorado

H. J. Goldsmid
Thermoelectric Refrigeration, 1964

G. T. Meaden
Electrical Resistance of Metals, 1965

E. S. R. Gopal
Specific Heats at Low Temperatures, 1966

M. G. Zabetakis
Safety with Cryogenic Fluids, 1967

D. H. Parkinson and B. E. Mulhall
The Generation of High Magnetic Fields, 1967

Volumes in preparation

J. L. Olsen and S. Gygax
Superconductivity for Engineers

A. J. Croft and P. V. E. McClintock
Cryogenic Laboratory Equipment

G. K. Gaulé
Superconductivity in Elements, Alloys, and Compounds

F. B. Canfield
Low-Temperature Phase Equilibria

W. E. Keller
Helium-3 and Helium-4

S. Ramaseshan
Low-Temperature Crystallography

P. E. Glaser and A. Wechsler
Cryogenic Insulation Systems

D. A. Wigley
The Mechanical Properties of Materials at Low Temperatures

S. A. Stern
Cryopumping

THE GENERATION OF HIGH MAGNETIC FIELDS

David H. Parkinson
Deputy Chief Scientific Officer
Head of Physics Group
Royal Radar Establishment, Malvern
and
Honorary Professor of Physics
University of Birmingham

and

Brian E. Mulhall
Senior Scientific Officer
Royal Radar Establishment, Malvern

Springer Science+Business Media, LLC

1967

Library of Congress Catalog Card Number 67-13568

ISBN 978-1-4899-6609-4   ISBN 978-1-4899-6612-4 (eBook)
DOI 10.1007/978-1-4899-6612-4
© 1967 Springer Science+Business Media New York
Originally published by Plenum Press in 1967.
Softcover reprint of the hardcover 1st edition 1967

Preface

The present book occupies a natural place in the growth of interest in high magnetic fields. Over the last few years a number of large international laboratories have been discussed or planned, and one, the National Magnet Laboratory at M.I.T., is now in full operation.

About five years ago we became involved in one such study. Particular attention was given to steady fields, of which we already had some experience. However, the field of interest covered the whole gamut of the techniques which might be employed and also the whole range of possible fields up to the extreme limits. It soon became evident that here was a mass of facts and experience which had not previously been collated. Thus the idea of this book was born.

Two particular problems have confronted us in the preparation. The first has been caused by the rapid advances in technology, especially those in superconductivity. Nevertheless, we feel that the present time is opportune for reviewing the situation. We have tried to present basic principles throughout as well as to create a useful source book for those interested in the subject. As with many technical problems, those involved here have no unique solutions, and compromises must be sought. We have tried to point out the conflicting factors in each case.

Second is the perennial problem of units. It is inevitable that a book of this type relies heavily on the literature. Some of the more recent papers have been written using rationalized mks units, many more using the cgs system, while those concerned with engineering practice are usually in the British (fps) system. We favor the mks system and have used it as widely as possible, particularly when discussing general principles; nevertheless, where appropriate, we have tried to facilitate comparison with the literature by using alternative systems. In all cases this is clearly stated.

Writing has been necessarily a spare-time occupation to be indulged in between experiments on the one hand and marshalling the resources of a solid state laboratory on the other. Much is based on first-hand experience and much on discussion with friends and colleagues, too numerous to be mentioned individually here. We freely acknowledge the value of these interchanges of ideas. In gaining our own experience we have relied heavily on our colleagues at the Royal Radar Establishment. In particular we are indebted to Messrs. A. G. Harris and F. Mansfield, on whom the responsibility has rested for the organisation and engineering of the High Field Laboratory and to Dr. J. Hulbert for many useful discussions concerned with pulsed fields and superconducting solenoids.

D. H. PARKINSON
B. E. MULHALL

September 1966

Contents

Notation

A	constant; area
B	magnetic flux density
C	capacitance; heat capacity C_c
E	energy level, gap, barrier
F	force; fluid friction constant
G	Fabry factor (Section 2.1.1); free energy
H	magnetic field, H_0, H_r, etc. at origin, r-component, etc.; H_c, H_{c_1}, H_{c_2} critical field of superconductors; heat transfer constant
I	electric current
J	electric current density
K	current density factor (Section 2.4.3)
L	inductance
M	intensity of magnetization
P	power density
Q	volume flow rate of fluid; $= H/I$ in superconductor
R	electrical resistance; radius in spherical coordinates
S	surface area
T	temperature, °K; T_c superconducting critical temperature
V	voltage
W	power
c	specific heat
d	diameter
e	electronic charge
f	frequency; constant
g	Landé splitting factor
h	Planck's constant ($\hbar = h/2\pi$); thermal exchange coefficient W/m²°C normalized field
j	normalized current density
k	Boltzmann constant; wave number; thermal conductivity

l	length
m	mass
n	integer; turns ratio; electron density
p	pressure; phase number
q	function of Q
r	radius or normalized radius in cylindrical coordinates; r_0 half solenoid bore
s	perimeter
t	sheet or lamina thickness; time; normalized temperature, T/T_0
v	velocity; volume
x, y	with z, Cartesian coordinates; variables
z	axial coordinate
α	normalized outer radius of coil; temperature coefficient of resistivity; α', stability parameter
β	normalized half length of coil
γ	electronic specific heat
δ	small distance; "skin" depth; δx small change in x
ϕ	angular coordinate; flux, ϕ_0 flux quantum
κ	Ginsburg–Landau parameter (Section 6.1.4)
λ	space factor; superconducting penetration depth
μ	permeability, $\mu_0 = 4\pi \times 10^{-7}$; viscosity; μ_0 Bohr magneton
ν	space factor; volume factor of coils
ω	angular frequency, $= 2\pi f$; Gorter–Casimir order parameter
ρ	electrical resistivity; radius in cylindrical coordinates (in place of r); density
σ	tensile stress
τ	time
η	Lagrangian multiplier; efficiency
θ	angle in spherical coordinates
ζ	axial coordinate (in place of z)
ξ	superconducting coherence length (Section 6.1.3)
Θ	Debye temperature
Δ	surface energy parameter; Δx large change in x
Ω	magnetostatic potential
Ψ	order parameter (Section 6.1.4)

THE GENERATION OF
HIGH MAGNETIC FIELDS

Chapter 1

Introduction and General Survey

1.0 INTRODUCTION

Throughout this book the term "high magnetic field" will refer to fields well in excess of 50 kOe (kilo-oersted), or in other words fields above those which can be reached by conventional iron-cored electro-magnets. First, it is necessary to state why it is worth writing about the generation of very strong magnetic fields and what they can be used for. Techniques for generating strong fields have been improving steadily over the years, the most recent step forward being the discovery of superconductors which have already been used to generate fields well above 100 kOe and which have the potential of going to considerably higher fields. Powerful fields, when once established in a superconduct-ing solenoid, can be maintained with no power dissipation other than that required to maintain the necessary low temperature, and high magnetic fields appear possible without the need for cooled solenoids dissipating megawatts of power or for massive iron-cored systems. Thus, technical processes demanding high fields which have hitherto been ruled out on economic grounds now become feasible, and in fundamental studies also the more extensive use of magnetic fields seems possible.

However, the situation is not quite as favorable as we might at first think, particularly when considering fields approximately 200 kOe and more. Regardless of the methods used to generate the fields, technical difficulties arise. There also appears to be a fundamental limitation on what can be expected of superconductors. Those systems which use solenoids dissipating very high powers (of the order of megawatts) have been called, light-heartedly, the brontosauri of magnetic field generation, but nevertheless for generating steady fields in excess of 200 kOe they are likely to remain the only practical pro-position for some time. As the situation appears now, superconducting solenoids, the high-powered "old-fashioned" systems, and pulsed coils for achieving the highest fields, are all mutually complementary, each one having applications in which it is the obvious choice. This is an interesting point in time at which to summarize and discuss the present technical situation. The subject is sufficiently wide for us to restrict the scope of this book to such a discussion.

1.1. APPLICATIONS OF HIGH FIELDS

1.1.0. The uses of high magnetic fields range from fundamental physical studies on the one hand, through applications to scientific instruments, and to applications in what may develop into industrial processes on the other. These uses can demand fields which are steady—lasting for minutes or hours at a time—or which may be of very short duration such as milliseconds or less. Some of the possible uses are given in the next paragraphs.

1.1.1. The principal need for high fields still arises in basic research, even though technological demands are increasing. In scientific applications, strong enough magnetic fields must always be capable of producing first-order effects in systems of particles (atoms, nuclei, or electrons) with magnetic moments, or in systems of moving particles carrying electric charge. In solid state physics there are a number of examples of the value of the use of magnetic fields as a tool.

One may be interested in localized electronic energy levels in crystals, as found in the salts of rare-earth and transition metal elements. In this case, magnetic fields can be used to establish the general nature of the lowest energy levels by the well-known paramagnetic resonance techniques. For this purpose comparatively low fields, of a few kilo-oersted, are all that is necessary. In other systems, such as paramagnetic ions and defects in ionic crystals, where there is strong interaction with the crystal lattice, the energy levels are broadened. Under these circumstances the splitting of a level by means of a magnetic field must be comparable at least with the level width in order to obtain useful information, and hence higher fields become necessary. The higher excited states of these localized systems in general also show broader levels and can be investigated by optical techniques—these in themselves, owing to resolution limits, also require high fields possibly ranging well above 100 kOe.

The use of paramagnetic rare-earth and transition metal salts in adiabatic demagnetization experiments for the generation of very low temperatures has long been a laboratory technique demanding high magnetic fields, for now the splitting of the energy levels, $g\mu_0 H$, must be appreciably less than the Boltzmann energy (kT), where

$T°K$ is the starting temperature of the demagnetization, usually around 1°K. Indeed, much of the more recent development of the orthodox water-cooled, high-powered solenoid systems has gone on in low-temperature laboratories. Temperatures in the millidegree range or less can be generated using electron spin systems, and temperatures of a few microdegrees have been produced using nuclear spin systems.

When the magnetic interactions between the localized electronic states are sufficiently large, the material becomes magnetically ordered at a low enough temperature—for example, ferromagnetics, anti-ferromagnetics, or ferrimagnetics. The degree of ordering as well as the nature of the electronic levels and of their interactions can be studied by carrying out, for example, resonance experiments in externally applied fields. To carry out antiferromagnetic resonance experiments in most materials, fields comparable with the internal field are necessary.

In metals and semiconductors the electronic energy levels are not localized, but are formed into quasi-continuous bands. The general structure of these bands can be studied by a number of techniques, most of which involve the use of magnetic fields. The application of a field quantizes the electronic motions perpendicular to the direction of the applied fields, and so gives rise to the formation of "Landau levels" in the electronic band structure. The quasi-continuous band structure is broken into a series of bands or "bundles" of levels separated from each other by an amount depending on the magnetic field. The separation is also inversely proportional to the effective electronic mass, m^*, within the band and so can be regarded as a measure of the relation between electronic energy and the wave vector for the band considered (in the simplest case, $E = \hbar^2 k^2 / 2m^*$). The Landau level separation can be measured directly by cyclotron resonance experiments where transitions take place between levels lying in the same band. Other experimental techniques involve the study of the de Haas–van Alphen effect, or the de Haas–Schubnikov effect. Where the effective mass is large, high fields, typically 100 kOe and more, are necessary to produce sufficient separation of the Landau levels. Another powerful technique which can be used with semiconductors is to study the optical absorption due to transitions between Landau levels in different electronic bands—e.g., valence to conduction.

Impurities in metals and semiconductors produce broadened Landau levels. To obtain sufficient separation under these conditions may require very high fields indeed. It has sometimes been stated that the use of high magnetic fields removes the necessity for extremely pure materials in this type of experiment, but this is a thoroughly misleading oversimplification of the situation. The ability to use very high fields in exploring the band structure of a solid can to some extent alleviate the necessity for very pure materials when carrying out a preliminary exploration. It can also enable one to obtain timely positive information as to the success of purification. However, in all work in which magnetic fields are used to sort out the electronic band structure or, indeed, localized electronic levels, high-purity single crystals are the first requisite; without them, much important fine structure could be lost.

In thinking of magnetic experiments with nuclei, the well-known nuclear resonance technique usually involves low fields, of a few kilo-oersted. There are other experiments which could be carried out using the nuclei of solids in much higher fields. For example, with a field sufficiently high, homogeneous, and steady, it should be possible to resolve the different resonance lines produced by the same atoms in different sites and environments in a disordered alloy. The effective field at the nucleus of an atom is about 500 kOe, and if fields of this order can be generated, a full investigation of the nuclear "hyperfine fields" is possible.

The recent discoveries concerning the use of superconductors to generate high magnetic fields have already been mentioned and will be dealt with later. In parallel, a much better understanding of superconductivity in general has arisen, and there is a much more clear recognition of what are now called type II superconductors. To be able to investigate their properties in full, and particularly those aspects which lead to or limit the continued existence of superconductivity in very high fields, high fields themselves are necessary, even though much of the fundamental understanding of these materials can result from studies in quite low fields.

1.1.2. On the technological front, magnetic fields have been extensively used in the containment of plasma in the general investigation of nuclear fusion processes—for example, in Zeta and in the mirror machines. In this type of application, much attention is now directed toward the use of superconductors for the generation of the fields required, which can range up to 100 kOe and more. Indeed, in all technological applications the emphasis must be on the use of hard superconductors for field generation. Thus the "magnetohydrodynamic" technique for the generation of electricity is made far more attractive with the use of superconductors to provide the necessary magnetic fields. Without them the energy consumed in generating the fields is likely to be greater than that generated, or at least so large that the process is not economically attractive. In the MHD generators proposed, a stream of extremely hot gas containing substances which ionize easily passes through a steady magnetic field—say, 50 kOe over a volume of several cubic meters. The ions separate in opposite directions and are collected on electrodes to give rise to a current in an external circuit.

These more massive industrial applications tend to overshadow other developments, particularly those in the field of electronic devices and scientific instruments. For example, a tunable far infrared detector based on indium antimonide has recently been demonstrated by Brown and Kimmitt (1963); of necessity the indium antimonide must be at a very low temperature (about 1.5°K) and in a magnetic field. The response is tunable by means of the field, going to shorter wavelengths at higher fields. For a maximum response at $100\,\mu$, a field of about 28 kOe is needed; but if steady fields of 200 kOe were available, the response would extend to between 10 and 15 μ.

Another example of a device requiring high magnetic fields is a generator of millimeter and submillimeter radiation described by Bott (1964) and known as the "teaser." In this device a beam of electrons travels in, and parallel to, a magnetic field. The electrons have a small transverse velocity and therefore execute transverse cyclotron orbits, which are characterized by quantum numbers and are analogous to Landau levels in a crystal. Owing to a small relativistic correction, the energy separations between orbits are not equal. The electron beam then passes into a region of very high magnetic field—e.g., 100 kOe—in a microwave cavity. Electrons can be stimulated to relax to orbits of lower quantum number with the emission of radiation. Power levels of about 1 W at wavelengths ranging from 4 mm to less than 1 mm have been reported.

1.1.3. The illustrations given above are sufficient to show that the interest in producing high magnetic fields stems from a wide front of research and technology. It is also worthy of note that many of the topics raised are concerned with those solids—viz., semiconductors, paramagnetics, ferromagnetics, and superconductors—from which most modern solid state devices are derived.

1.2. TECHNIQUES OF MAGNETIC FIELD GENERATION

1.2.0. The techniques of high-field generation can be divided into three groups, though all have the common feature that large electric currents are driven through conductors wound into solenoids, the field being generated along the axis of the coil. These groups are:

1. The use of normal conductors, usually copper, working near room temperature and cooled with water or occasionally some other liquid.

2. The use of normal conductors, but cooled with a cryogenic liquid such as nitrogen or hydrogen.

3. The use of superconductors working at liquid helium temperatures.

All three techniques can be used to generate steady fields, and the first two can also be used for pulsed fields of short duration.

1.2.1. It is useful to compare the different techniques in terms of what can be done with high-powered normal solenoids. It is easy to show (cf. Section 2.1.1) that the axial field produced at the center of a solenoid of uniform filling factor λ is given by the expression:

$$H_0 = G\sqrt{W\lambda/r_0\rho} \qquad (1.1)$$

where W is the power dissipated in the coil, r_0 is the inner radius, ρ is the resistivity of the conductors, and G is a constant for a particular magnet geometry. For the majority of solid state experiments it is possible to work with r_0 near 2 cm. For practicable constructions, Equation (1.1) reduces to the useful approximate expression:

$$H_0 \sim 110\sqrt{W/r_0} \quad [\text{kOe}] \qquad (1.2)$$

with W in megawatts and r_0 in centimeters. Thus, to generate a field of 110 kOe in a coil of 2 cm inner radius requires a power of about 2 MW. The problems of designing and building such coils are now fairly well known, and the power level is not exorbitantly high. On going to higher fields, however, various problems arise. The first is that of ensuring adequate cooling in the solenoid; ideally the current should be concentrated at the center close to where the field is wanted, but this requires a large cooling area and hence many cooling channels there, which is inconsistent with the need for maximum current density. The limitations imposed by heating considerations result in coils which are a little larger and less efficient than may otherwise have been the case.

Fields can then be generated which will produce stresses in the coil which exceed the mechanical strength of the conductor material; for normal constructions this is near 250 kOe. The condition requiring that the stresses shall be kept below the limits imposed by the conductors must be satisfied, for example, by suitably adjusting the current distribution in the coil, which once more causes the solenoid to become larger and considerably less efficient. Nevertheless, provided that enough power is at hand, steady fields in the region of 400 kOe appear to be feasible. At these high fields the power costs become very high indeed, particularly because the field generated for a given power is considerably less than that indicated by Equation (1.2).

1.2.2. As a means of overcoming heavy power consumption it is tempting to consider the use of conductors cooled with liquid nitrogen or liquid hydrogen. Thus, the resistivity of pure copper 77°K is about 1/8, and at 20°K may be as low as 1/1000 of that at room temperature. With hydrogen cooling one can envisage electrical power consumption measured in tens of kilowatts instead of megawatts. However, energy must be expended to produce the cryogenic liquid, and it is obvious that the operational conditions are important. If the system is to be used continuously, the refrigeration power can be greater than the electrical power of the conventional system. If, on the other hand, the system is to be used only infrequently, cryogenic cooling can be an advantage, for heavy electricity costs may be avoided by producing the coolant at a steady, but slow, rate and storing it until such time as it is released at a very high rate through the solenoid.

Cryogenic solenoids suffer from cooling problems similar to those of water-cooled solenoids, and in any event, the mechanical strength problems are very little changed.

1.2.3. Superconducting solenoids at present have only generated fields up to about 130 kOe, although the problems peculiar to them will no doubt be overcome in time. Nevertheless there appears to be a thermodynamic limit to the field which they can support (Clogston, 1962), which is probably near 250 kOe. Meanwhile the outstanding problem has been to produce solenoids in which the critical current-carrying characteristic approaches that given by short sample tests. The degradation in performance of a superconducting material when wound into a solenoid can be by a factor as high as 5. However, a means of overcoming this difficulty and of building large solenoids which are inherently stable and in which the wire behaves with its short-sample characteristics has now been discovered.

1.2.4. Turning now to the production of magnetic fields of short duration, typically a few milliseconds, we find that the plant needed—usually of the type in which a capacitor is discharged through the coil—is very inexpensive compared with that for orthodox steady-field systems. The short duration of the field imposes the condition that in any experiment all natural and instrumental time constants must be very short. This is a difficult condition to meet, for example, in magneto-optical experiments, particularly those in the far infrared. Nevertheless, the pulse techniques are a very useful complement to the steady-field systems. Very high fields indeed can be generated in this way, and it is the only way at present of approaching 1000 kOe or more. Although the electromagnetic forces in the solenoid may be well beyond the mechanical strength of the coil, it can remain intact, provided that the impulse is short enough. Thus, it is comparatively easy to produce fields near 400 kOe in simple coils of about 7-mm bore in pulses with a duration of about 1 msec. Higher fields can be produced for shorter periods, such as 700 kOe in pulses lasting for a few tens of micro-seconds. Ultimately, however, the magnetic forces become so great that the coil disintegrates during the first pulse. An implosion technique has been used to produce fields as high as 10^7 Oe for about 2 μsec, but of course both coil and experimental sample are destroyed in one shot.

1.2.5. In discussing the generation of high fields in detail, the question of solenoid design must occupy first place, with problems of heat dissipation and mechanical strength predominating. Even with superconductors this is true, except that here the heat dissipation which is most troublesome is that occurring should the coil inadvertently become normal. With all techniques the power plant or current supply cannot be ignored, and in fact this and the control plant together may be much more expensive than the solenoid system itself.

1.3. HISTORICAL REVIEW

The history of the development of techniques for generating strong magnetic fields is of interest. Even before 1900 Fabry (1898) had investigated theoretically the possibility of generating high fields using air-cored solenoids. Deslandres and Perot (1914) built and successfully tested a solenoid cooled with chilled paraffin, which could generate fields of about 50 kOe. At about this time also a very large iron-cored electromagnet was being designed and built at Bellevue, near Paris, soon to be followed by another at Uppsala in Sweden. Both were capable of producing fields in the region of 40 to 50 kOe in useful volumes. The prime interest at that time was to be able to study the Zeeman effect in simple atomic systems, for it was then that the rapid advances in atomic physics were in progress.

In the period from 1925 to 1940 the beginnings of some of the present developments were already visible—at least, with the wisdom of hindsight they can be seen now. The idea of using demagnetization techniques with paramagnetic salts to generate very low temperatures was suggested independently by Giauque (1926) and Debye (1926) and for a period the main drive toward generating stronger steady fields came from the very few low-temperature laboratories then in existence. In general, the higher the field, the lower the temperature which could be reached, provided that the right paramagnetic salts could be found. In 1929 and 1930 the first investigations of the lead–bismuth alloys, now known to be hard superconductors, were in progress, although their potential was not realized until much later. It was in this period also that Kapitza began the first work producing fields of 200 kOe and higher, using pulse techniques. In the years 1935 to 1940 Bitter extended the earlier mathematical analyses, and in a series of experiments at the Massachusetts Institute of Technology he completed the foundations of modern orthodox techniques for the generation of steady fields.

Since 1945 the general upsurge of activity in solid state physics has brought a renewed interest in using magnetic fields as tools to help in elucidating the electronic and magnetic structure of crystals. The need to be able to split energy levels or resolve Landau levels which are naturally broad has added incentive to pushing the magnitude of both steady and pulsed fields as high as possible. Today steady fields of over 200 kOe have been produced, with the prospect of even greater fields not too far distant.

1.4. CENTERS OF HIGH-FIELD WORK

1.4.0. It is worthwhile to consider the laboratories throughout the world in which high magnetic fields are used as research tools. Pulsed fields, using capacitor discharge techniques, are used in a large number of laboratories, because such systems, though limited in application, are inexpensive to set up. Superconducting solenoids are beginning to be used extensively, but at the time of writing, the highest fields available in which experiments can be done with them are about 100 kOe.

1.4.1. Higher steady fields require more expensive installations. In England there are three laboratories in which high-powered, water-cooled solenoids are used:

1. Clarendon Laboratory, Oxford, where there is a motor generator set of 2 MW rating and where fields of near 120 kOe can be generated. Research is concentrated on nuclear alignment experiments, nuclear cooling, and the behavior of paramagnetic ions in high fields.

2. Cavendish Laboratory, Cambridge, where there is a 2-MW silicon rectifier–transformer plant and fields about 90 kOe are available. Research is concentrated on studies of the Fermi surfaces of metals.

3. Royal Radar Establishment, Malvern, which has a battery of lead–acid cells capable of yielding up to 3.5 MW as well as a 2.6-MW rectifier-transformer system. Fields up to 150 kOe are available. Research is concentrated on the behavior of hard superconductors and on magneto-optical studies in semiconductors.

Elsewhere in Europe there are at present three installations:

1. Kamerlingh Onnes Laboratory, Leiden, with a transformer–mercury arc rectifier equipment capable of delivering 4 MW at 1000 V. Fields to about 90 kOe are known to have been generated.

2. Polish Academy of Science, Wroclaw, where there is a motor–generator set capable of delivering 3.5 MW at 500 V.

3. Laboratoire d'Electrostatique et de Physique du Metal, Grenoble, where fields in the region of 100 kOe have been generated.

There are no published references to high steady-field facilities in the USSR. In the United States there are a number of laboratories in operation:

1. National Magnet Laboratory at the Massachusetts Institute of Technology, where there are generators capable of 8 MW continuously or 12 MW for 15 min. This laboratory has only recently opened, having

absorbed the old, much smaller, M.I.T. magnet laboratory set up by Bitter. Fields up to 120 kOe have been available for some time, and now 220 kOe has been exceeded in a solenoid of 4 cm bore. Research covers a wide area of solid state physics, including the study of semiconductors, metals, and paramagnetic ions in solids, using a wide variety of techniques.

2. Naval Research Laboratories, Washington, where there is a generator rated at 2 MW, but capable of delivering up to 3 MW for 10 min. Fields up to about 150 kOe are available, with which work is done on the behavior of metals and semiconductors in high fields.

3. Bell Telephone Laboratories, Murray Hill, where a small motor generator set can give up to 1.6 MW for periods of a few minutes. The maximum field is about 100 kOe in an iron-clad solenoid; research is concentrated on the behavior of semiconductors, metals, and paramagnetic ions in solids.

4. Department of Chemistry, University of California, Berkeley. The motor generator set is rated at 6 MW, 700 V. Fields up to 100 kOe have been generated, and have been used almost exclusively in problems connected with the entropy changes occurring in the excitation of paramagnetic ions in solids at low temperatures.

5. Lewis Research Center, NASA, Cleveland, where there is a homopolar generator rated continuously at 1.8 MW, but which can be overrun to about 3 MW for 1 min. The fields reached are around 100 kOe, with which magnetoresistance studies in metals have been attempted.

6. Los Alamos Scientific Laboratory, which has a cryogenic solenoid installation using liquid hydrogen as coolant and consuming 50 kW of electrical power at 5 V, and where fields up to 80 kOe have been generated. Research on hard superconducting materials is in progress.

7. National Bureau of Standards, Boulder, Colorado, where the techniques of cryogenic solenoids using liquid hydrogen and generating fields of up to 70 kOe have been investigated.

8. Oak Ridge National Laboratory, Oak Ridge, Tennessee, where fields near 80 kOe have been generated.

One more installation, should also be noted:

Research Institute for Iron, Steel, and Other Metals, Tohoku University, Sendai, Japan. Using a mercury-arc rectifier plant capable of a continuous direct-current output of 4 MW, fields of 120 kOe have been generated.

There are a number of laboratories in which orthodox high-field installations are being seriously prepared, such as that at the University of Pennsylvania and that at Graduate Center of the Southwest, Dallas, Texas. It is known, too, that in France a large national laboratory is proposed and actively being planned, to be located at Grenoble. Power levels comparable to those at the American National Magnet Laboratory would be available. A laboratory is also being planned at Nijmegen in Holland.

It is apparent from the list above how large a proportion of the effort in this field is concentrated in the United States. It is also noticeable how much of the effort is at the 2–4 MW power level. This is because the fields which can thereby be generated are large enough to be useful, as Equation (1.2) shows, and yet the costs and technical difficulties are not too great. Steady fields must be pushed to worthwhile higher levels; 250 kOe and beyond has been already lightly indicated, but the costs and the technical difficulties are great. It is one of our purposes to discuss this extension from the known techniques of the 100 to the 250 kOe region, and then extrapolate into the currently more speculative region of steady fields around 400 kOe.

Chapter 2

Solenoid Analysis

2.0. INTRODUCTION

As outlined in Chapter 1, to generate powerful steady magnetic fields one has the choice between dissipating large amounts of power in windings of normal conductors, or using quantities of expensive superconducting materials to form the windings. The requirements are such that it is interesting to determine the coil configuration which minimizes either the power or the quantity of superconductor needed, and a theoretical discussion of coil design is devoted largely to these points.

The simplest optimization was considered as long ago as 1873 by Clerk Maxwell, in connection with galvanometer coils. Then followed the work of Fabry (1898, 1910), by whose name the geometry factor G of Equation (1.1) is usually known, on a more practical form of coil. Kapitza (1927), Cockcroft (1928), and Bitter (1936) in a classic paper, considered the problems which arise at higher field levels—viz., those of the electromagnetic forces generated in the coil, and of distributing a large volume of cooling passages through the windings.

In the case of superconducting coils the cooling considerations are no longer relevant (at least not in this form); instead the current-carrying capacity of the wire becomes less at higher field levels. To minimize the quantity of wire needed, all parts should obviously carry the maximum possible current. These topics are deferred until Chapter 7.

With the expanding interest in the application of magnetic fields, the optimization problem is more general than merely maximizing the field for a given expenditure of electrical power. Often, for example, a maximum field nonuniformity will be specified, thus forcing one away from a classical optimum configuration. A different generalization concerns coils employing low-temperature windings, in which the power needed to pump the coolant around may become comparable to the electrical power; such a case has been discussed by Kronauer (1962). It is, however, not useful to overgeneralize, because the coil design is ultimately influenced by many factors which cannot be given mathematical form—for example, ease of manufacture and reliability—and the rewards to be won become progressively smaller.

2.1. ANALYSIS OF OPTIMIZATION

2.1.0. In the analysis it is convenient to use normalized quantities. For the dimensional variables (cylindrical coordinates r, z) the normalizing constant is half the solenoid bore r_0, while for the current density it is J_0, the density at radius r_0 and at $z = 0$. λ is a space factor, expressing the proportion of the coil volume which is effectively filled with conducting material; it will generally be treated as a continuous, slowly varying function of position as a more convenient alternative to evaluating integrals over just the volume occupied by conducting material.

There are two possible ways of defining current density, one in which the current in an elementary volume is averaged over the entire cross section, and the other in which the average is only over the effective conducting cross section. We shall use the latter, and denote it by j; the average current density is then λj.

2.1.1. In general, coils will have cylindrical symmetry. They will have a plane of symmetry normal to the axis, and the current may be considered to have only an azimuthal component. The field produced at the origin, and the power dissipated in an elementary

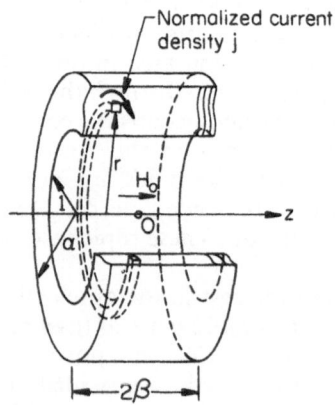

Fig. 2.1. Coil notation

hoop as sketched in Figure 2.1 can be written down, and the expressions integrated over the entire cross section to give the central field, H_0, and the electrical power dissipated, W:

$$H_0 = \frac{r_0 J_0}{2} \iint \frac{\lambda j r^2}{(r^2 + z^2)^{3/2}} \, dr \, dz \tag{2.1}$$

$$W = 2\pi r_0^3 J_0^2 \iint \rho \lambda r j^2 \, dr \, dz \tag{2.2}$$

In those cases where the space factor and the resistivity ρ are independent of position, they may be taken outside the integrals. J_0 may then be eliminated between Equations (2.1) and (2.2) to yield

$$H_0 = G(W\lambda/r_0\rho)^{1/2} \tag{2.3}$$

where

$$G = \left(8\pi \iint r j^2 \, dr \, dz \right)^{-1/2} \iint \frac{j r^2 \, dr \, dz}{(r^2 + z^2)^{3/2}} \tag{2.4}$$

G is the Fabry factor, the generally used figure of merit of a coil first introduced by Fabry. It is seen to depend only on the form and relative dimensions of the coil (these determining the limits for the integrals) and on the way the current is distributed.

2.1.2. We shall now consider optimization in the simplest case, that of a conventional coil of rectangular section when both the resistivity and space factor are constant. Following Bitter (1936) the function $H_0 + \eta W$ is maximized, subject to the condition that W be constant. η is a Lagrangian (undetermined) multiplier. The variables are the functional form of the current density j and the normalized dimensions. We shall further suppose that the current varies in the radial direction only (since this is a useful practical situation), so that Equations (2.1) and (2.2) may first be integrated with respect to z between the limits $-\beta$ and $+\beta$. Thus,

$$H_0 = \frac{r_0 J_0 \lambda}{2} \int_1^\alpha \frac{2\beta j \, dr}{(r^2 + \beta^2)^{1/2}} \tag{2.5}$$

$$W = 2\pi r_0^3 J_0^2 \lambda \rho \int_1^\alpha 2\beta r j^2 \, dr \tag{2.6}$$

For a stationary value the variation of $H_0 + \eta W$ is zero, and hence

$$\frac{r_0 J_0 \lambda}{2} \frac{2\beta}{(r^2 + \beta^2)^{1/2}} + (\eta 2\pi r_0^3 J_0^2 \lambda \rho)(4\beta r j) = 0 \tag{2.7}$$

Because J_0 is defined such that $j \equiv 1$ at $r = 1$, the constants in Equation

(2.7) are such that j may be written:

$$j = \frac{1}{r}\sqrt{\frac{1 + \beta^2}{r^2 + \beta^2}} \tag{2.8}$$

On substituting this expression for j into Equation (2.4), or into Equations (2.5) and (2.6), it will be seen that G is given by

$$G = \sqrt{\frac{1}{8\pi\beta}\log\left(\frac{\alpha^2(1 + \beta^2)}{\alpha^2 + \beta^2}\right)} \tag{2.9}$$

This increases with α, and for $\alpha \to \infty$ has a maximum near $\beta = 2$. It is apparent that the stationary point thus found does represent an optimum configuration.

2.1.3. Analyses similar to that above have been carried out for a large number of geometries, and useful collections of results are to be found in papers by Bitter (1936), by Montgomery and Terrell (1961), and by Gauster (1960). They are collected together for reference in the Appendix to this chapter.

From these examples it is apparent that there is an upper limit to the value of the Fabry factor, attained in an infinite coil with the optimum current distribution $j = r/(r^2 + z^2)^{3/2}$ and with only a sphere of radius r_0 left as experimental space. The upper limit is then 0.23 (or 0.289 in cgs units). Gauster has shown that, whatever the shape of a coil or assembly of coils in which the current distribution is of this optimum form, the Fabry factor is always improved by an increase in the coil cross section.

For more practical current distributions, and finite coils, lower Fabry factors are found, in the region of 0.15 (0.19 in cgs units). This is not necessarily a very serious loss, for the current densities are lower than in a fully optimized solenoid, and thus the heat transfer and mechanical stress problems are somewhat less severe.

2.2. EXTENSION OF ANALYSIS TO INCLUDE COOLING

2.2.0. When considering practical solenoids to dissipate powers of more than a few hundred kilowatts, the analysis of the previous section is only a first approximation. For the second approximation the space factor and resistivity can no longer be assumed constant.

The resistivity of the conductors is a function both of temperature and of magnetic field. Magnetoresistive effects need only be considered in coils cooled to below about 30°K; they are not easily analyzed, because there is no simple expression for the field at an arbitrary point in a solenoid, and they will therefore not be discussed at this stage.

The thermal variation of resistivity can be quite adequately expressed by a linear law, except again at temperatures of the order of 30°K and lower. We shall assume

$$\rho = \rho_0(1 + \alpha\Delta T) \tag{2.10}$$

where ΔT is the difference between the temperature at which the resistivity is ρ, and that at which it is ρ_0. For copper, ρ_0 at 20°C is about 1.75×10^{-8} Ω-m, and α 4×10^{-3} per °C. In typical magnets ΔT is about 60°C, so that the working resistivity is about 2.25×10^{-8} Ω-m. Other resistivity data are given in Section 5.2.1.

There are two contributions to the space factor, one coming from the necessary electrical insulation between the turns of the coil, and the other coming from the cooling passages which must pass through the windings. It is usually only the latter contribution which varies significantly with position; we shall let the local fraction of volume effectively occupied by coolant be v. The insulators are made as thin as possible, and the fraction of volume they occupy, $1 - \lambda_i$, only varies if the conductor cross section is varied in order to realize some desired current distribution. The space factor is thus $\lambda_i - v$. It is sometimes more convenient to define λ_i as the ratio of conductor thickness to that of conductor plus insulator, and in this case the space factor is $\lambda_i(1 - v)$. This produces just the same results as considering the resistivity of the conductors to be ρ/λ_i and decreasing the current density by a factor λ_i to allow for the increased section, and is useful when considering Bitter solenoids.

2.2.1. The distribution of cooling voids and the current distribution are intimately connected, for in very high-power solenoids the heat must be removed where it is generated. To do otherwise would require very large temperature differentials in the magnet, which no currently used insulators could withstand—even if high enough heat transfer rates could be achieved at the cooled faces. In the case of very wide conductors, such as in a Bitter magnet, the current would tend to shun the hottest zones, modifying the current distribution in an unpredictable manner. Thus the cooling void fraction v is put proportional to the local power dissipation:

$$v = \frac{j^2 J_0^2 \rho \lambda}{f}$$

$$= \frac{j^2 J_0^2 \rho \lambda_i(1 - v)}{f} \tag{2.11}$$

f is the heat which can be removed per unit volume of cooling void, and is usually proportional to the temperature differential between the coolant and the conductors.

Using Equation (2.11), λ can be eliminated from Equations (2.1) and (2.2), the resulting expressions for field and power being

$$H_0 = \frac{r_0 J_0 \lambda_i}{2} \iint \frac{jr^2 \, dr \, dz}{(r^2 + z^2)^{3/2}(1 + j^2 A)} \tag{2.12}$$

and

$$W = 2\pi r_0^3 J_0^2 \rho \lambda_i \iint \frac{rj^2}{(1 + j^2 A)} \, dr \, dz \tag{2.13}$$

A has been written in place of $J_0^2 \rho \lambda_i / f$. Both A and ρ are functions of temperature, and a complete optimization would include this. In practice, however, the temperature is limited to about 150°C maximum for normal operation by three factors: by the insulators used, by unstable heat transfer conditions arising if the conductor temperature is much above the boiling point of the coolant, and by some conductor materials, in particular pure copper, already tending to anneal and become less stress- or creep-resistant. If, then, some temperature can be assumed, the optimum form of j can in principle be determined by the procedure outlined in Section 2.1.1. However, the solution is now more complicated, and depends on the magnitude of the current, or in other words, on the field level. This is illustrated by an example. Again the current is supposed to vary only in the radial direction. Equations (2.12) and (2.13) are integrated with respect to z, and then the function $H_0 + \eta W$ differentiated with respect to j as before. For a stationary value this differential is zero; the multiplier η is determined as before by using the condition that at the inner edge of the winding ($r = 1$) j equals 1. The expression for j thus becomes

$$Aj^2 + \frac{(1 - A)r(r^2 + \beta^2)^{1/2}j}{(1 + \beta^2)^{1/2}} - 1 = 0 \tag{2.14}$$

When A is zero, this is precisely the result derived before. For some other values of A the results are shown in Figure 2.2.

To determine the optimum solenoid dimensions it is necessary to substitute the value of j defined by Equation (2.14) back into the field and power integrals, Equations (2.12) and (2.13), and hence calculate the Fabry factor or its derivatives with respect to α and β. Essentially this has been done in a few special cases, but no useful sets of results are known.

For small values of A, such as might be appropriate to a 2-MW solenoid, it can be seen that the optimum current density function is not very different from $1/r$. This is, of course, Bitter's classic result.

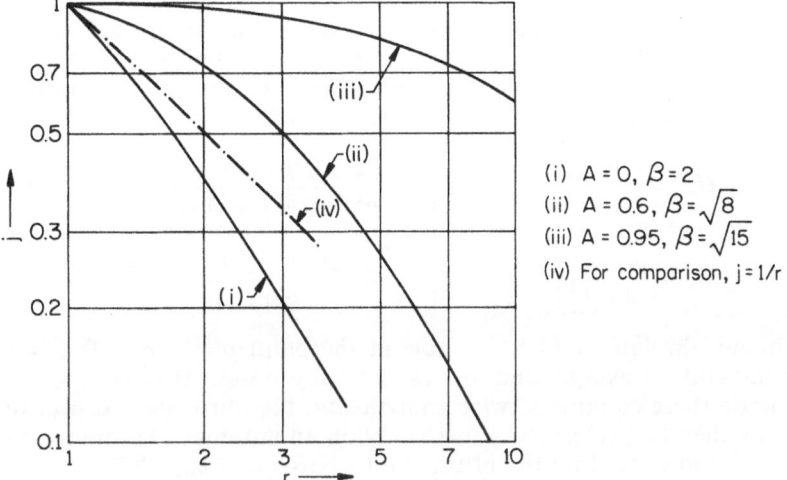

Fig. 2.2. Optimized current distribution functions.

2.3. FIELD AWAY FROM THE ORIGIN

2.3.0. In many aspects of solenoid design or application it is desired to know the field at points other than the origin. Thus, in the complete analysis of the electromagnetic forces in a solenoid, or in the complete optimization of a superconducting coil, the field at points inside the windings is required. In the more sophisticated experiments needing high fields it is usual to specify a maximum field inhomogeneity which can be tolerated in the experimental space.

Although this question has been investigated, using a variety of approaches, explicit and useful analytic expressions cannot be derived. However, tables of field functions, compiled with the aid of automatic computers, have been published (Brown *et al.*, 1963; Alexander and Downing, 1959). At its simplest, the winding is considered as a number of current-carrying hoops, the fields due to which can be calculated quite easily. At off-axis points, elliptic integrals are involved. To achieve good accuracy a large number of hoops must be taken, and the computation is rather slow. More elegant is the method of Garrett (1962), which is based on expressing the field as a series of spherical harmonics. It suffers from the disadvantage that the spheres of convergence of the series may not extend into the windings; this can be overcome, but to calculate the field at points inside the winding, much more sophisticated programming is needed.

In this discussion we shall limit ourselves to an outline of the problem.

2.3.1. A perfectly general formal expression for the field components at an arbitrary point $(r, 0, z)$ can be written down, by considering the winding divided into elementary hoops and integrating, just as was done in determining the central field. We find

$$H_z = \frac{J_0 r_0}{4\pi} \iiint_0^{2\pi} \frac{j(\rho - r \cos \phi)\rho \, d\phi \, d\zeta \, d\rho}{[(z - \zeta)^2 + \rho^2 - 2\rho r \cos \phi + r^2]^{3/2}} \quad (2.15a)$$

$$H_r = \frac{J_0 r_0}{4\pi} \iiint_0^{2\pi} \frac{j(z - \zeta)\rho \cos \phi \, d\phi \, d\zeta \, d\rho}{[(z - \zeta)^2 + \rho^2 - 2\rho r \cos \phi + r^2]^{3/2}} \quad (2.15b)$$

Although the integrand has a pole at the point $\rho = r$, $\phi = 0$, $\zeta = z$ the integrals converge, and are valid at any point. It is possible to integrate these equations twice analytically; the third, with respect to ϕ, may then be performed numerically on an automatic computer, as was done in compiling the NASA tables (Brown *et al.*, 1963).

2.3.2. Equation (2.15a) is of further use in determining the field homogeneity near the origin. The expression for H_z is expanded formally in a power series in z along the axis—so that r is now zero—as follows:

$$H_z = \sum_{n=0}^{\infty} H_n z^n \quad (2.16)$$

where

$$H_0 = \text{field at the origin}$$

$$H_n = \frac{1}{n!} \frac{\partial^n H_z}{\partial z^n}\bigg|_{z=0} \quad (2.17)$$

Usually the solenoid will be symmetrical about the origin, and then

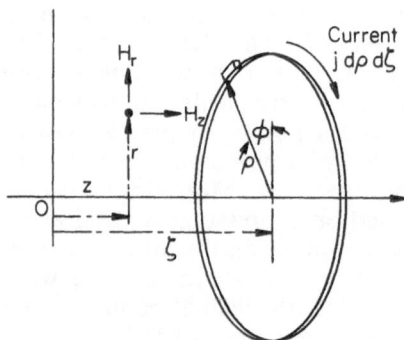

Fig. 2.3. Coordinate notation for off-axis
field calculations.

H_n is zero for odd n. The axial expansion is extended to nonaxial points by the method suggested by Garret (1951). Outside the windings, and in particular around the origin, the field is the gradient of a scalar potential, Ω say, which satisfies Laplace's equation. Hence, in spherical polar coordinates, R, θ—no azimuthal coordinate is required because of the symmetry—Ω takes the form

$$\Omega = \sum_{n=0}^{\infty} A_n R^n P_n(\cos\theta) \qquad (2.18)$$

the A_n being as yet undetermined constants. The axial field is expressed in terms of the spherical components, H_R, H_θ, by

$$H_z = \sum_{n=1}^{\infty} nAR^{n-1}P_{n-1}(\cos\theta) \qquad (2.19)$$

On the axis all the $P_n(\cos\theta)$ have modulus unity, such that $R^n P_n(\cos\theta)$ becomes just z^n. The coefficients in Equation (2.19) are then obtained by comparison with Equation (2.17), which gives

$$(n+1)A_{n+1} = H_n \qquad (2.20)$$

The corresponding expression for the radial field is

$$H_r = \sum_{n=1}^{\infty} -A_n R^{n-1} \sin\theta\, P'_{n-1}(\cos\theta) \qquad (2.21)$$

In the report by Montgomery and Terrell (1961), values of the first few coefficients H_n/H_0 are tabulated for the useful cases $j = 1$ and $j = 1/r$, for a range of coil dimensions. Table 2.I shows expressions for one or two of the lower order coefficients in these two cases. In practice the higher terms are not significant, for one does not have

Table 2.I. Field Coefficients[*]

	$j = 1/r$	$j = 1$
H_0	$2\log\left\{\alpha \cdot \dfrac{\beta + \sqrt{1+\beta^2}}{\beta + \sqrt{\alpha^2+\beta^2}}\right\}$	$2\beta\log\left\{\dfrac{\alpha + \sqrt{\alpha^2+\beta^2}}{1 + \sqrt{1+\beta^2}}\right\}$
H_2	$\beta\left\{\dfrac{1}{(\alpha^2+\beta^2)^{3/2}} - \dfrac{1}{(1+\beta^2)^{3/2}}\right\}$	$\dfrac{1}{\beta}\left\{\dfrac{1}{(1+\beta^2)^{3/2}} - \dfrac{\alpha^3}{(\alpha^2+\beta^2)^{3/2}}\right\}$
H_4	$\dfrac{\beta}{4}\left\{\dfrac{2\beta^2-3\alpha^2}{(\alpha^2+\beta^2)^{7/2}} - \dfrac{2\beta^2-3}{(1+\beta^2)^{7/2}}\right\}$	$\dfrac{1}{12\beta^3}\left\{\dfrac{20\beta^4+7\beta^2+2}{(1+\beta^2)^{7/2}} - \dfrac{(20\beta^4+7\beta^2\alpha^2+2\alpha^4)\alpha^3}{(\alpha^2+\beta^2)^{7/2}}\right\}$

[*] Omitted is a factor $J_0 r_0/2$ if mks units are used, or $\pi J_0 r_0/5$ in cgs units.

sufficient control over the field configuration for them to be worth considering.

The problem of determining the coil dimensions when the maximum field homogeneity tolerable is specified is not so elegantly treated. The obvious condition to impose is that the first few coefficients, H_n, should vanish, but this gives highly nonlinear equations to be solved. Franzen (1962) has considered the case—hardly valid for high power solenoids—where the dimensions are so small that the equations can be expanded as a Taylor series, and thus linearized. More generally this condition can be combined as a subsidiary condition in the optimization problems already discussed. In practice, given good computing facilities, the task of designing coil shapes for which the H_n vanish is not too difficult; one proposes a geometry which may be expected to be suitable, and allows a sufficient number of dimensions to be variables. For example, in a single coil H_2 is negative. By splitting the coil and separating the halves, as a Helmholtz pair, a configuration can be found for which H_2 vanishes, although for typical coils the separation is unfortunately too small for useful radial access to the origin. Furthermore, by suitable choice of overall length it is also possible to maximize the efficiency (or Fabry factor). By replacing the separating gap by a region with thicker conductors an additional variable, namely, the ratio of the current densities in the two halves, is created, which allows H_4 to be made zero. The procedure described has proved useful for Bitter magnets, but other configurations suitable for different constructions are readily proposed.

An alternative approach to designing a uniform-field coil is considered by Montgomery and Terrell (1961). The coil is divided up into a number of narrow coils placed side by side (Figure 2.4).

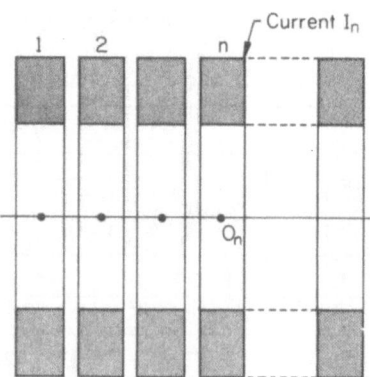

Fig. 2.4. Coil subdivision for homogeneity
calculations.

The current in each is then to be determined such that the resultant field is the same at the center of each coil. The field contributions from each coil are the product of the current and a factor which is calculated by solving an equation such as (2.15) for $r = 0$. Once these factors are calculated, the problem reduces to one of solving a number of linear simultaneous equations for the currents. The variation of the field along the axis is obviously reduced by taking a larger number of coils to start with; it can in any case be estimated by computing the field midway between two of the coils. Equations (2.19) and (2.21) show the degree of radial uniformity which may be expected once the axial field is made uniform.

2.4. ELECTROMAGNETIC FORCES

2.4.0. The electromagnetic forces on an isolated current loop (Figure 2.5) are easily calculated. The components of magnetic induction are B_r, B_z, which equal μH_r, μH_z, respectively, and the current in the loop at radius r is $\lambda J_0 jr_0^2 \delta r \, \delta z$. The forces can be resolved into an axial component δF_z and a radial component δF_r, which may be written as the product of current, induction, and length of element, thus

$$\delta F_z = \lambda J_0 jr_0^2 \, \delta r \, \delta z B_r rr_0 \, \delta\phi \qquad (2.22)$$

$$\delta F_r = \lambda J_0 jr_0^2 \, \delta r \, \delta z \, B_z rr_0 \, \delta\phi \qquad (2.23)$$

If we further assume that the radial forces are contained entirely by a tensile stress, σ, in the conductors (or $\lambda\sigma$ acting over the full cross

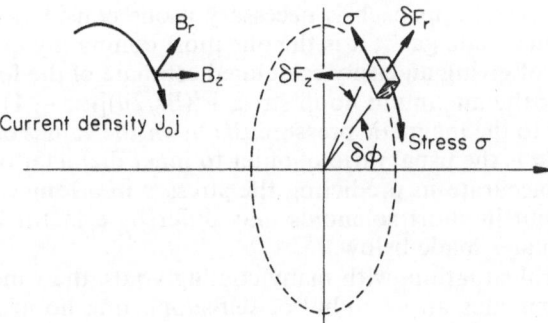

Fig. 2.5. Forces on isolated current element.

section of the hoop) we find

$$\delta F_r = \lambda \sigma r_0^2 \, \delta r \, \delta z \, \delta \phi \qquad (2.24)$$

and hence

$$\sigma = J_0 j r_0 r B_z \qquad (2.25)$$

2.4.1. To evaluate further the electromagnetic forces on each hoop in a winding requires a knowledge of the field distribution. Such an analysis was carried out by Kapitza (1927), and in greater detail by Cockcroft (1928); they, however, expressed the field as a function of current and the mutual inductance between each hoop and the solenoid. The mutual inductance was calculated using a method due to Butterworth. An alternative, far simpler approach is to regard the forces as originating from the pressure, $B_0 H_0/2$, of a "magnetic fluid" inside the solenoid bore, or to divide the winding into shells with the magnetic fluid creating a pressure differential across each shell (Montgomery, 1963).

The value of each approach depends on how one regards the stress as distributed in the windings. Since one can calculate the electromagnetic force on each element, one can, in principle, perform a stress analysis on the whole winding; but in practice of course this is extremely laborious. It has, however, been done (Daniels, 1953), in the relatively simple case of a stack of narrow "pancake" coils where it was possible to stress in the radial direction only, ignoring the axial forces. Similar calculations have also been made by Léon (1964). More usually the problem is simplified in one of three ways: by allowing each hoop to support its own radial forces, by regarding the winding as a homogeneous solid which is then stressed as a thick-walled pressure vessel, or by regarding the windings as plastic and transmitting all the forces to an outer shell. The last approach was adopted by Kapitza, probably justifiably at the field levels at which he was working. The pressure vessel approach is necessary if one considers the forces as due to a "magnetic gas"; it is the one most commonly quoted, and has the merit of giving an easily calculated estimate of the forces. The expression for the maximum hoop stress is $(B_0^2/2\mu)[(\alpha^2 + 1)/(\alpha^2 - 1)]$, which is close to the magnetic pressure $B_0^2/2\mu$ for the values of α usually encountered (α is the usual ratio of outer to inner diameter of the coil). This is most accurate in predicting the stresses in a long solenoid of moderate α, but in short solenoids may differ by a factor 2 or more from the estimates made below.

In any real situation, with many cooling voids, the winding most nearly approximates an assembly of self-supporting hoops. Because the hoops in a winding are not closed, friction between turns is necessary—this point has been discussed by Giauque and Lyon (1960) in connection with a particular strip-wound coil. The arrangement is as sketched in Figure 2.6; for a helix of small pitch the normal force between turns in $p\,2\pi r r_0^2\,\delta r$, while the shear force at the interface is $\sigma r_0^2\,\delta r\,\delta z$. The coefficient of friction must obviously be greater than a critical value, μ_{crit}, given by

$$\mu_{crit} = \sigma\,\delta z/2\pi r p \qquad (2.26)$$

In typical coils the pressure p generated by the axial forces is sufficient to ensure that the frictional forces are adequate over most of the center part of the winding, but additional support is needed at the ends, and possibly also at the outside edge where the electromagnetic forces change direction.

2.4.2. A rough estimate of the pressures generated can be made, for the axial forces can only be supported against adjacent turns and must ultimately appear as compressive forces in the center plane of the solenoid. The shear strength between radially adjacent hoops is assumed to be zero, so that δF_z acting on the area $r r_0^2\,\delta r\,\delta\varphi$ is equivalent to an axial pressure, δp, where

$$\delta p = \lambda J_0 j r_0 B_r\,\delta z \qquad (2.27)$$

δp is integrated in the axial direction from z up to β to yield the pressure acting at point r, z. Since div $\mathbf{B}$ is zero, we have the relation

$$\frac{1}{r}\cdot\frac{\partial}{\partial r}(rB_r) = -\frac{\partial B_z}{\partial z} \qquad (2.28)$$

which may be integrated

$$rB_r = \int_0^r -r\frac{\partial B_z}{\partial z}\,dr \qquad (2.29)$$

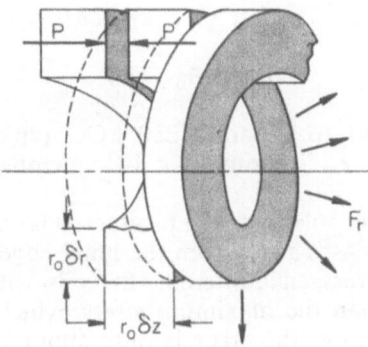

Fig. 2.6. Forces on helical coil.

Often j is independent of z, so that on substituting the expression (2.29) for B_r into the expression (2.27) for δp the integration may be performed to yield

$$p = \frac{J_0 r_0 j \lambda}{r} \int_0^r -r[B_z]_z^\beta \, dr \tag{2.30}$$

To the approximation already implied it is sufficient to consider B_z as constant up to $r = 1$, though it does in fact increase to a few percent above B_o at the inner edge of the winding. The pressure at $r = 1$ is thus of order $j J_0 r_0 (B_0 - B_\beta)$, where B_β is the axial flux density at the end of the winding.

2.4.3. To estimate the greatest tensile stresses which are set up in a winding we consider Equation (2.25). The stress at the inner edge of the winding will be written σ_0, and thus

$$\sigma_0 = J_0 r_0 B_0 \tag{2.31}$$

B_0 and the current density J_0 are related by Equation (2.1):

$$B_0 = \frac{\mu_0 r_0 J_0}{2} \iint \frac{\lambda j r^2}{(r^2 + z^2)^{3/2}} \, dr \, dz \tag{2.1}$$

A new geometry factor, K, can be defined as the integral

$$K = \frac{1}{2} \iint \frac{j r^2}{(r^2 + z^2)^{3/2}} \, dr \, dz \tag{2.32}$$

and Equation (2.1) thus becomes

$$B_0 = \mu_0 r_0 \lambda J_0 K \tag{2.33}$$

Equation (2.31) for σ_0 can now be rewritten:

$$\sigma_0 = B_0^2 / \mu_0 K \lambda \tag{2.34}$$

In a Bitter solenoid to generate 250 kOe, typical values may be $\lambda K = 1.25$, and thus σ_0 is about 40×10^7 newtons/m², or 4000 atmospheres.

Because in Bitter solenoids the product jr is constant and the flux density always decreases away from the inner edge of the winding, σ_0 is the maximum stress encountered. In coils with uniform current density σ_0 is less than the maximum stress, which now occurs some way inside the winding; the error is only about 20% for small coils ($\alpha \sim 2$ or 3), but may be 100% and more for deep windings.

2.4.4. The stresses at 250 kOe are already at the ultimate tensile strength of pure copper, and if higher fields are to be reached, this barrier must be circumvented. Alloys of higher strength than copper may be used, in theory perhaps to double the maximum possible field for pure copper. However, the resistivity of such alloys is great, lead- to much increased power requirements and greatly intensified cooling problems.

An alternative approach is to increase the current factor K—in other words, to decrease the current density in the windings. To do this the winding volume is increased, while at the same time the Fabry factor is decreased, as is illustrated in Figure 2.7. The decrease in

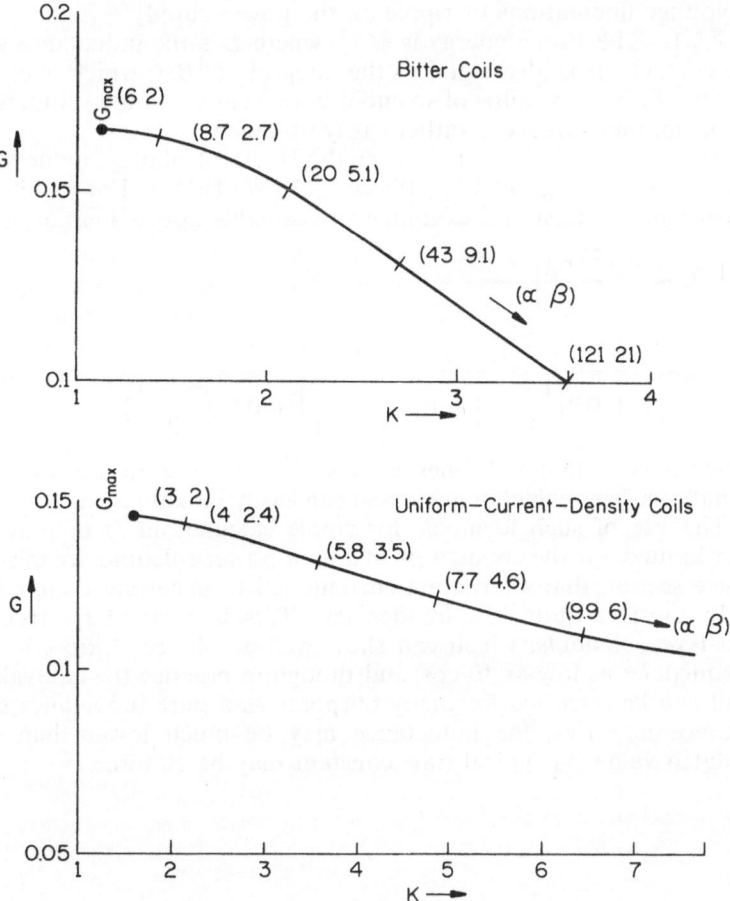

Fig. 2.7. Fabry factor G vs. current density factor K.

Fabry factor is to some extent offset by the increase in space factor resulting from the easier cooling requirements.

A third approach is to depart from the traditional cylindrical coil, and instead to seek configurations in which the field and current vectors in the windings are parallel.

These topics are discussed further in Chapter 9.

2.5. INDUCTANCE CALCULATIONS

2.5.0. The inductance of a solenoid is of interest in considerations both of the stored energy to be dissipated after a fault and of smoothing any voltage fluctuations or ripple on the power supply.

2.5.1.. The stored energy is $\frac{1}{2}LI^2$, where L is the inductance and I the current. It is also equal to the integral of $\frac{1}{2}BH$, which is of the order of $\frac{1}{2}B_0H_0 \times$ volume of solenoid bore. Thus a rough estimate of the solenoid inductance is rather easily obtained.

More accurate formulae or methods of calculating inductance are well known—e.g., Welsby (1960) or Grover (1946). For a uniform current coil of n turns, for example, a reasonable approximation is:

$$L \sim \frac{\mu_0 n^2 \pi r_0 (1 + \alpha)^2}{8\beta}$$

$$\times \frac{1}{1 + 0.9\left(\frac{1 + \alpha}{4\beta}\right) + 0.64\left(\frac{\alpha - 1}{\alpha + 1}\right) + 0.84\left(\frac{\alpha - 1}{2\beta}\right) + \dots} \quad (2.35)$$

Montgomery (1963) publishes curves showing the results of such calculations, from which inductances can easily be estimated.

The use of such formulae for ripple calculations is in practice rather limited, for the conductors in a high-power solenoid are of such massive section that alternating currents fail to penetrate them completely, even at quite low frequencies. This is particularly true of Bitter-type solenoids, which can show well-developed "skin effects" at frequencies as low as 10 cps, and though in practice the equivalent circuit can be regarded for many purposes as a pure inductance and resistance in series, the inductance may be much lower than the calculated value. A typical time constant may be 20 msec.

Appendix

ANALYSIS OF SOLENOID GEOMETRIES

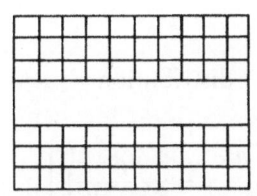

$$j = 1$$

$$G = \left[\frac{\beta}{2\pi(\alpha^2 - 1)}\right]^{1/2} \log\left[\frac{\alpha + (\alpha^2 + \beta^2)^{1/2}}{1 + (1 + \beta^2)^{1/2}}\right]$$

$$K = \beta \log\left[\frac{\alpha + (\alpha^2 + \beta^2)^{1/2}}{1 + (1 + \beta^2)^{1/2}}\right]$$

$$G_{max} = 0.142 \text{ at } \alpha = 3, \beta = 2$$

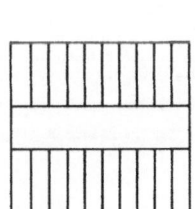

$$j = 1/r \qquad \text{``Bitter'' distribution}$$

$$G = \frac{1}{(4\pi\beta \log \alpha)^{1/2}} \log\left[\alpha\left(\frac{\beta + \sqrt{1 + \beta^2}}{\beta + \sqrt{\alpha^2 + \beta^2}}\right)\right]$$

$$K = \log\left[\alpha\left(\frac{\beta + \sqrt{1 + \beta^2}}{\beta + \sqrt{\alpha^2 + \beta^2}}\right)\right]$$

$$G_{max} = 0.166 \text{ at } \alpha = 6, \beta = 2$$

$$j = \frac{1}{r}\sqrt{\frac{1 + \beta^2}{r^2 + \beta^2}}$$

Best possible function of r only for rectangular coils

$$G = \frac{1}{(8\pi\beta)^{1/2}}\left[\log\frac{\alpha^2(1 + \beta^2)}{\alpha^2 + \beta^2}\right]^{1/2}$$

$$K = \frac{(1 + \beta^2)^{1/2}}{2\beta} \log\frac{\alpha^2(1 + \beta^2)}{\alpha^2 + \beta^2}$$

$$G_{max} = 0.179 \text{ at } \alpha \to \infty, \beta = 2$$

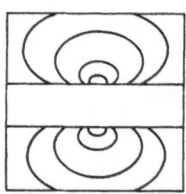

$$j = \frac{r}{(r^2 + z^2)^{3/2}} \qquad \text{optimum ``Kelvin'' distribution}$$

$$G = \sqrt{\left[\frac{1}{32\pi}\left(3\tan^{-1}\beta - \frac{3}{\alpha}\tan^{-1}\frac{\beta}{\alpha} + \frac{\beta}{1+\beta^2} - \frac{\beta}{\alpha^2 + \beta^2}\right)\right]}$$

$$K = \frac{1}{8}\left(3\tan^{-1}\beta - \frac{3}{\alpha}\tan^{-1}\frac{\beta}{\alpha} + \frac{\beta'}{1+\beta^2} - \frac{\beta}{\alpha^2 + \beta^2}\right)$$

$$G_{\max} = 0.217 \text{ at } \alpha, \beta \to \infty$$

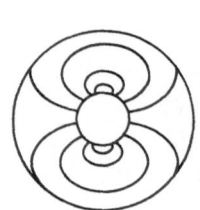

$$j = \frac{r}{(r^2 + z^2)^{3/2}}$$

$$G = \sqrt{\frac{\alpha - 1}{6\pi\alpha}}$$

$$K = \frac{2}{3}\left(\frac{\alpha - 1}{\alpha}\right)$$

$$G_{\max} = 0.230 \text{ as } \alpha \to \infty$$

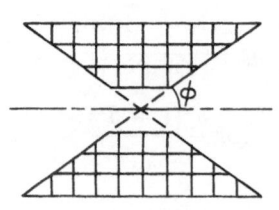

$$j = 1$$

$$G = \frac{\alpha - 1}{\sqrt{\alpha^3 - 1}}\left[\frac{3k}{4\pi(1 + k^2)}\right]^{1/2}$$

$$K = \frac{\alpha - 1}{(1 + k^2)^{1/2}} \ *$$

$$G_{\max} = 0.137 \text{ at } \phi = 45°, \ \alpha = 2.7$$

---

* $k = \tan\phi$.

$$j = \frac{1}{r}$$

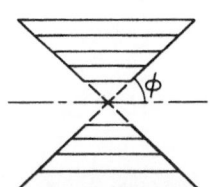

$$G = \sqrt{\frac{k}{1 + k^2}} \frac{\log \alpha}{\sqrt{\alpha - 1}} \frac{1}{\sqrt{4\pi}}$$

$$K = \frac{\log \alpha}{\sqrt{1 + k^2}} *$$

$$G_{\text{max}} = 0.160 \text{ at } \phi = 45°, \alpha = 4.5$$

$$j = \frac{1}{r^2}$$ Optimum function of r only for tapered coils

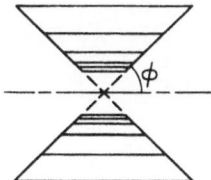

$$G = \sqrt{\frac{\alpha - 1}{\alpha} \cdot \frac{k}{1 + k^2} \cdot \frac{1}{4\pi}}$$

$$K = \frac{\alpha - 1}{\alpha} \cdot \frac{1}{\sqrt{1 + k^2}} *$$

$$G_{\text{max}} = 0.199 \text{ at } \phi = 45°, \alpha \to \infty$$

$$j = \frac{r}{(r^2 + z^2)^{3/2}}$$

$$G = \sqrt{\frac{\alpha - 1}{4\pi\alpha}\left[\frac{3}{8}\left(\frac{\pi}{2} - \phi\right) + \frac{1}{4}\sin 2\phi - \frac{1}{32}\sin 4\phi\right]}$$

$$K = \frac{\alpha - 1}{\alpha}\left[\frac{3}{8}\left(\frac{\pi}{2} - \phi\right) + \frac{1}{4}\sin 2\phi - \frac{1}{32}\sin 4\phi\right]$$

$$G_{\text{max}} = 0.217 \text{ as } \phi \to 0, \alpha \to \infty$$

---

* $k = \tan \phi$.

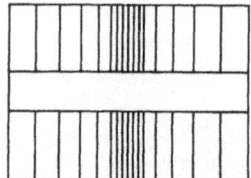

$$j = \frac{1}{r}\left\{\frac{1}{\sqrt{1+z^2}} - \frac{1}{\sqrt{\alpha^2+z^2}}\right\}\frac{\alpha}{\alpha-1}$$

"Gaume" distribution, optimum of form $\frac{1}{r}f(z)$

$$G = \sqrt{\frac{1}{4\pi \log \alpha}\left[\tan^{-1}\beta + \frac{1}{\alpha}\tan^{-1}\frac{\beta}{\alpha} - \frac{2}{\alpha}F(\phi,k)\right]}$$

$$K = \frac{\alpha}{\alpha-1}\left(\tan^{-1}\beta + \frac{1}{\alpha}\tan^{-1}\frac{\beta}{\alpha} - \frac{2}{\alpha}F(\phi,k)\right)^*$$

$G_{\text{max}} = 0.184$ at $\alpha = \beta = 8$

---

* $F(\phi,k)$ = Elliptic integral of first kind; $\phi = \tan^{-1}\beta$; $k = \sqrt{1 - 1/\alpha^2}$.

Chapter 3

Cooling

3.0. INTRODUCTION

In the design of magnets to generate steady high fields, one of the major problems is that of the removal of heat from the windings. This is already severe in conventional magnets at the 100 kOe or 2 MW level, and at higher fields considerable difficulties arise in making a winding which is adequately cooled and yet mechanically strong. The related problem in superconducting coils is that of preventing the conductor from going normal accidentally, and protecting the coil and other apparatus if it should do so. This, however, is more properly dealt with when discussing superconducting coils in Chapter 7. The present chapter is therefore devoted to a discussion of intensive cooling in conventional solenoids, and of the properties of some of the coolants which might be used.

3.1. THEORY OF COOLING

3.1.0. High-power solenoids are cooled by passing fluid through many small passages in the windings. To achieve the best possible heat transfer, the coolant is brought into direct contact with the conductors where the heat is generated. The passages can assume a variety of forms, as will be described later; for the time being, let us consider a passage of length l, and of circular cross section of diameter d.

If it is necessary to consider passages of noncircular section, the analysis will be valid with d replaced by an equivalent "hydraulic" diameter d_h, which is usually defined in terms of the ratio of cross-sectional area S to wetted perimeter s thus:

$$d_h = 4S/s \tag{3.1}$$

However, this can only be satisfactorily justified for passages not very different from circular or square.

3.1.1. The analysis which follows is based on equations quoted by McAdams (1954). They are correlations of the results of a wide variety of experiments, and can be used with confidence within the limits of the original data. For the turbulent flow of an incompressible fluid,

$$p = p_e + p_f \tag{3.2}$$

$$p_e = \frac{3\rho v^2}{4} \tag{3.3}$$

$$p_f = \frac{0.092\rho l v^2}{d}(N_{Re})^{-0.2}$$

$$= Fl\frac{v^{1.8}}{d^{1.2}} \tag{3.4}$$

N_{Re}, which equals $dv\rho/\mu$, is the Reynolds number for the flow. The fluid has velocity v, density ρ, and viscosity μ. p is the total pressure drop, which comprises a frictional component p_f, and an entrance loss p_e. In calculating this entrance loss, it is assumed that the changes of section are abrupt and that the fluid outside the small passage has negligible velocity; formulae are available for other circumstances.* Equation (3.4) is valid for the range of Reynolds number from about 5000 to 200,000, which is adequate for most solenoid design applications; for use outside this range, other formulae and graphs have been published.*

For the heat transfer, the appropriate equations are:

$$h = 0.023\, k(\rho/\mu)^{0.8}(N_{Pr})^{0.4}v^{0.8}/d^{0.2}$$

$$= Hv^{0.8}/d^{0.2} \tag{3.5}$$

where h is the power transmitted across unit area of the conductor–coolant interface for unit temperature differential; k is the thermal conductivity of the fluid, and N_{Pr} its Prandtl number, which equals μ/k times the specific heat, c; all are evaluated at the bulk temperature, because further refinement will rarely be warranted in solenoid design work. Thus, if water is considered and the variation of viscosity with temperature is neglected, an error of less than 15% results in the above formulae. This is of the same order as the uncertainty in the values of the coefficients in the formulae, and far less than that caused by slight roughness or fouling of the flow passages.

In reality, the cooling passages often are far from smooth, and the formulae (3.4) and (3.5) are no longer valid. However, it is only the

---

* For example, the Pipe Friction Manual of the New York Hydraulic Institute.

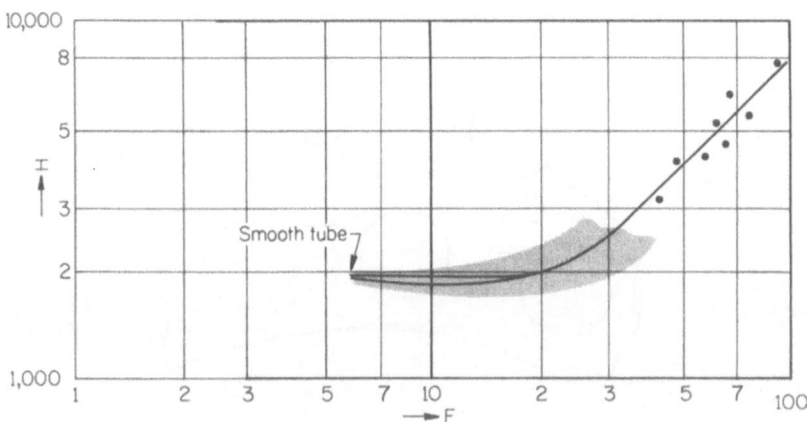

Fig. 3.1. Correlation between pipe friction and heat transfer.

coefficients H and F which differ considerably from the predicted values, rather than the functional dependence of heat transfer or pressure drop on coolant velocity and hydraulic diameter. Theoretically for water at 20°C, F should be about 5.8 and H 1900 in mks units; in practice, F can reach 50 or 100, depending on the solenoid construction, and the corresponding value of H around 4000. There is no clear correlation between pipe friction and heat transfer efficiency, because the spread in experimental results is considerable, but in general H increases with F, as illustrated in Figure 3.1.

3.1.2. Using the formulae (3.4) and (3.5), it is possible to investigate how the temperature in a solenoid varies with cooling passage size. For simplicity each cooling passage will be considered to be carrying the heat away from a prism of area S_0. The effective "waste space" factor v will be greater than S/S_0, by a factor which depends on the distortion of the current flow around the cooling passage. We can write

$$v = KS/S_0 \tag{3.6}$$

For example, for circular holes such as might be made in a Bitter disk, the lines of current flow are as sketched in Figure 3.2. For a regular, infinite grid, the potential problem can be solved formally, or an analog can be made in resistance paper. It is found that the effective area of the holes is about twice their geometric area. If the holes can be arranged to cause no distortion of the lines of current flow, K can obviously assume the minimum possible value of 1—which is approached by oval or rectangular holes suitably oriented.

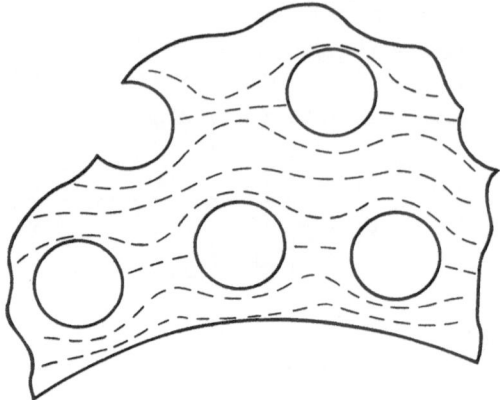

Fig. 3.2. Lines of current flow around cooling
holes.

It will be assumed that the coolant is at a fixed pressure. This is
reasonable in the case of greatest current interest, namely that of water-
cooled magnets, where even at the highest practical pressures the pump
power is still quite small compared with the electrical power. In other
cases the starting assumptions are different, and the problem in general
much more complicated. Often the entrance loss is small compared
with the frictional loss (e.g., in a Bitter solenoid p_e may be only one-
tenth of the total pressure loss), and in what follows it will be neglected.
The power density in the conductors is P, and thus the heat to be
removed is $P(S_0 - S)l$. Writing the temperature rise in the coolant
ΔT_1 and the temperature difference between coolant and conductors
ΔT_2, it can be seen that

$$\Delta T_1 = \frac{P(S_0 - S)l}{Qc\rho} \tag{3.7}$$

$$\Delta T_2 = \frac{P(S_0 - S)l}{hsl} \tag{3.8}$$

Q is the volume of coolant per second, and equals Sv. The velocity and
volume of coolant are now eliminated, to yield

$$\Delta T = \Delta T_1 + \Delta T_2$$
$$= \frac{P(S_0 - S)}{S}\left[\left(\frac{Fl}{p_f}\right)^{5/9}\frac{1}{c\rho d^{2/3}} + \left(\frac{Fl}{p_f}\right)^{4/9}\left(\frac{d^{2/3}}{4Hl}\right)\right] \tag{3.9}$$

This can be rewritten:

$$\Delta T = \frac{K - v}{v} \frac{Pl}{2} \left(\frac{F}{c\rho p_f H}\right)^{1/2} \left[\left(\frac{d}{d_0}\right)^{2/3} + \left(\frac{d_0}{d}\right)^{2/3}\right] \qquad (3.10)$$

where

$$d_0 = \left(\frac{4Hl^{10/9} \cdot F^{1/9}}{c\rho p_f^{1/9}}\right)^{3/4} \qquad (3.11)$$

This clearly reaches a minimum when d equals d_0, at which the temperature is ΔT_0:

$$\Delta T_0 = \frac{(K - v)}{v} Pl \left(\frac{F}{Hcp_f\rho}\right)^{1/2} \qquad (3.12)$$

3.1.3. Typical values of the various parameters for water are collected in Table 3.I, and substituting these in Equation (3.11) it can be seen that the optimum hole diameter is in the region of one or two millimeters. The corresponding number of holes may be one or two thousand, which represents a formidable manufacturing problem. Figures (3.3) and (3.4) show the calculated temperature rise of both coolant and conductors as a function of distance through the solenoid and of hole diameter. For values of d in the range $2d_0/3$ to $3d_0/2$, the temperature reached at the exit is roughly constant. As d is decreased below d_0, the maximum temperature is approached only near the exit end, while the mean temperature goes through a minimum of about $0.7 \Delta T_0$ near $d_0/2$. For larger values of d, the conductors are at a much more uniform temperature, and the mean temperature is higher. Thus,

Table 3.I. Properties of Water at 30°C

Density	1000 kg/m³
Specific heat, c	4180 joules/kg
Thermal conductivity, k	0.592 joule/m-°C
Viscosity, μ	0.001
Coil length	0.2 m
Frictional pressure	
Drop, p_f	6 × 10⁵ newtons/m²
	(6 atm)
Theoretical value of	
F	5.8
H	1900
d_0	0.873 mm
Typical value in Bitter solenoid of	
F	58
H	3800
d_0	1.78 mm

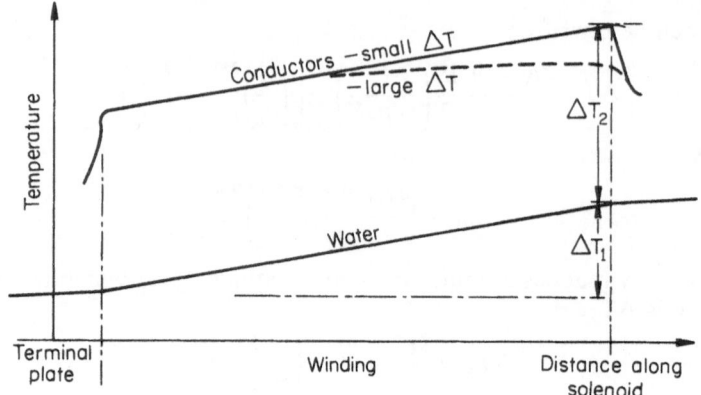

Fig. 3.3. Temperature distribution in solenoid.

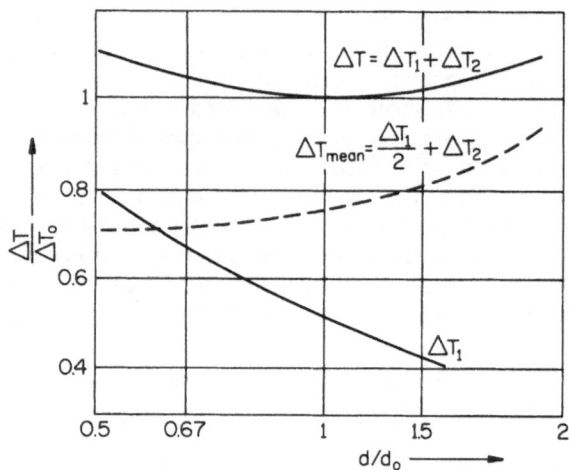

Fig. 3.4. Variation of temperatures with cooling hole
diameter.

if d is about $3d_0/2$, the exit temperature is only 4% greater than the
optimum, but the mean temperature is about 15% greater. Even so,
the increase in electrical power needed rises to only about 4% at full
power, which is a small price to pay for the easing of the manufacturing
difficulties. From the point of view of magnet operation this may not
be so desirable: if a magnet fails, it often starts to break down at the
hottest point, and it is reassuring to have a large length of well-cooled
conductor to limit the fault current and inhibit the spread of the fault.

It also seems likely, as we shall discuss below, that the maximum allowable heat flux is greater for the smaller holes.

3.2. NUCLEATE BOILING HEAT TRANSFER

3.2.0. As the power density in the conductors is raised, the mechanism of heat transfer into the coolant changes. A point finally comes at which conditions become unstable and the temperature rises dramatically and often disastrously.

The general form which the temperature differential takes as a function of heat flux is as sketched in Figure 3.5. The analysis of the section above is applicable to the region A–B, where the temperature of the heated surface never appreciably exceeds the local boiling point of the coolant. If the temperature is increased beyond this point, a region of much increased heat transfer coefficient is encountered until C, at which conditions finally become unstable and the temperature rises suddenly to D, where it will often be above the melting point of the metal. In the region of enhanced heat transfer, vapor bubbles are forming at the heated surface and recondensing in the bulk fluid. It is thought that the expanding bubbles increase the turbulence and thus promote better mixing of the heated fluid in the boundary layer with the cool bulk. At the "burnout" point C, the bubbles fail to recondense quickly enough, and an insulating film forms at the surface. In the confined space of a magnet-cooling passage, the metal reaches its melting point, while the steam expands violently. The writers have observed magnets in which this is presumed to have happened, where molten copper was projected back several feet against the coolant flow!

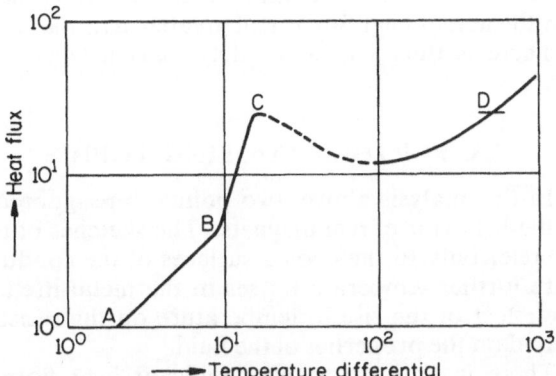

Fig. 3.5. Heat transfer characteristic of metal–coolant interface.

3.2.1. There is a considerable body of literature on the subject of nucleate boiling, but for our purposes the results do not differ significantly from those reported by McAdams (1954, Chapter 14). The fluid considered is water, for it is on this that most work has been done, but the general principles remain true for other liquids.

When the heat flux is plotted against the difference between the temperature of the heated surface and the boiling point of the liquid, a curve is obtained which is virtually independent of the velocity of the fluid and the degree of subcooling. The general shape of the curves for higher pressures—i.e., for higher boiling points—is similar but the temperature differentials are reduced; however, the difference over the range up to 15 atm is small. The critical heat flux, at which the heat transfer becomes unstable, increases with pressure to a maximum at about one-third of the critical pressure, or about 70 atm for water, at which it may be three or four times that at 1 atm.

The critical heat flux also varies with velocity, and with degree of subcooling. Various correlations are reported; all agree in that the critical flux varies as v^n, where n lies between $\frac{1}{3}$ and $\frac{1}{2}$. The flux also increases with the subcooling, again as a power lower than the first. These apply to water under conditions similar to those occurring in a solenoid, although unfortunately little is known about the way the heat flux depends on surface conditions, particularly roughness of the type found in a Bitter solenoid. However, in the magnet referred to above, breakdown occurred at a flux of about $1 \, kW/cm^2$, which is consistent with a predicted flux of 1 to $1.5 \, kW/cm^2$.

If the length of the solenoid is increased or the diameter of the cooling holes decreased (the space factor of course remaining unchanged), the predominant effect is a reduction in heat flux. Although the water velocity is decreased and its temperature increased, the reduction in the actual heat flux is still greater than that in the critical heat flux. There is thus a greater safety margin when operating at high powers.

3.3. FURTHER COOLING TOPICS

3.3.0. In the analysis above, two points were ignored which are relevant to the behavior of real magnets. The sketches of temperature distribution refer only to the cooled surfaces of the conductor, but in fact there are further temperature rises in the metal itself. Also neglected is the effect of the rise in temperature on the resistivity of the conductors and on the properties of the fluid.

3.3.1. There is unlikely to be significant heat flow across the insulators—their thermal conductivity is typically one-thousandth that of copper—and the question of temperature distribution usually reduces to a simple problem in one of two dimensions. In a Bitter solenoid, for example, we can consider a cooling tube drawing heat from a cylindrical volume round it (Figure 3.6). The only variation to be calculated is in the radial direction, and the equation to be solved is thus

$$\frac{k}{r}\frac{d}{dr}\left(r\frac{d\Delta T_r}{dr}\right) = -P \tag{3.13}$$

Hence it may be shown that the temperature at radius r is ΔT_r above that at r_0, where

$$\Delta T_r = \frac{P}{2k}\left[r_0^2 \log\left(\frac{r}{r_1}\right) - \frac{r^2 - r_1^2}{2}\right] \tag{3.14}$$

The distribution is sketched in Figure 3.6b, including also the temperature between conductor and coolant for comparison. The parameters are typical of a Bitter solenoid to dissipate about 3.0 MW— length 0.2 m, 620 cooling holes, of about 2.5 mm diameter. It does, of course, depend on the spacing of the holes, and therefore even though the holes are distributed to remove the heat uniformly, the mean temperature will not be uniform; in fact, the outside will be a few degrees hotter than the inside.

The fact that there are temperature gradients in the conductors implies that the resistivity is not constant; this will have, however, only a second-order effect on the current distribution, and has been neglected.

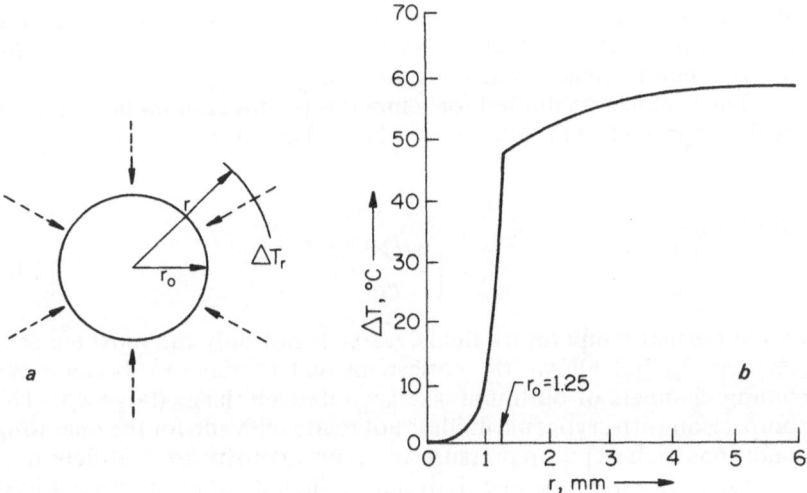

Fig. 3.6. Temperature gradients near cooling hole.

3.3.2. A further point concerns the temperature differential between the conductors and the coolant, which in the full curve of Figure 3.3 was shown as constant over the whole length of the magnet. In practice, the heat transfer efficiency varies, due to both the variation of coolant properties with temperature and the approach to nucleate boiling as discussed above. The heat flux at the exit end is also greater, due to the increase in resistivity. The overall effect clearly depends on the magnitude of the temperature differential; thus for a well-cooled strip-wound magnet Giauque and Lyon (1960) report a relatively small net effect, while for a Bitter magnet working at high heat flux we may expect the dotted curve of Figure 3.3.

3.4. COOLANT FLUIDS

3.4.1. The properties of most liquids which might be considered for a magnet cooling system are listed in Table 3.II. For water, kerosene, and glycol the properties are quoted for a temperature of about 30°C, which is a typical operating condition, while for the other liquids the temperature assumed is the normal boiling point.

It is certainly most convenient to pump or draw the cryogenic liquids from a reservoir under about 1 atm pressure. However, a higher heat transfer rate and a more stable operating temperature can be achieved by suppressing actual boiling in the magnet and thus preventing two-phase flow—say, by subcooling the reservoir or by providing overpressure even at the magnet outlet.

3.4.2. If boiling does occur, it is unlikely to result in a disastrous burnout, for any cooling failure allows the resistance to rise and limit the current (Laquer, 1962). Nevertheless, the field fluctuations this causes would be unacceptable for most purposes.

The coefficients quoted for temperature and cooling hole size are the fluid parameters in Equations (3.11) and (3.12), viz.,

$$T_c = \sqrt{F/Hc\rho} \tag{3.15}$$

$$d_c = \left(\frac{4HF^{1/9}}{c\rho} \right)^{3/4} \tag{3.16}$$

Of the normal temperature fluids, water is not only the most efficient (smallest T_c) but allows the easiest magnet fabrication, because the cooling channels of optimum size are relatively large (large d_c). The comparison with cryogenic fluids is not really relevant, for the operating conditions such as pump pressure and power density are so different.

3.4.3. It would also be desirable to include data on critical heat flux, but insufficient information is available. For pool boiling in water

Table 3.II. Properties of Coolant Fluids

	Liquid H₂*	Liquid D₂*	Liquid Ne§	Liquid N₂*	Water†	Kerosene‡	Glycol†
Boiling point at 1 atm, °K	20.4	23.5	27.2	77.4	373	>450	
Critical pressure, atm	12.8	16.3	27	33.5	220		
Critical temperature, °K	33.3	38.2	44.5	126	647		
Temperature, °K	20	23	27	77	293	303	293
Density, kg/m³	71	160	1200	804	1000	797	900
Specific heat, J/kg-°C	9300	5860	1800	2040	4180	1840	2400
Thermal conductivity, W/m-°C	0.118	0.133	0.115	0.139	0.592	0.148	0.265
Viscosity ($\times 10^4$)	0.138	0.298	1.96	1.6	10.2	16.3	240
$T_c(\times 10^5)$	2.6	2.9	4.5	4.5	2.7	9.5	14
$d_c(\times 10^2)$	1.4	1.4	1.1	1.2	1	0.7	0.37

Sources of constants:

* Scott, 1959.
† McAdams, 1954.
‡ Giauque and Lyon, 1960.
§ Huth, 1962; Löchtermann, 1963; Bewilogua and Mahn, 1963.

the critical flux is about 120 W/cm^2, while in nitrogen and hydrogen it may reach 15 W/cm^2 and 8 W/cm^2 respectively (most of the available experimental data on cryogenic liquids are collected in an article by Brentari and Smith, 1965). In narrow channels with natural convection flow, the critical flux may be much less (Sydoriak and Roberts, 1957); on the other hand, under forced convection and with subcooling, higher fluxes can be attained, perhaps by a factor 10 or so in water.

Chapter 4

Solenoid Construction and Instrumentation

4.0. INTRODUCTION

The theory of Chapters 2 and 3 deals with those aspects of magnet design which are amenable to analysis. The results are, nevertheless, more qualitative than quantitative, and can only indicate the directions in which the best design is to be found, as opposed to being a complete set of design formulae. Much depends also on the form of construction to be adopted, which is greatly influenced by individual preferences and by the engineering resources available. In what follows, some of the constructions so far adopted will be described and an indication of their limitations given.

4.1. GENERAL CONSTRUCTIONAL CONSIDERATIONS

Certain of the solenoid parameters are determined from the start by the nature of the power plant. Thus the voltage and current ratings are known, and the solenoid is designed to have a definite resistance at its working temperature. Likewise the maximum coolant pressure and flow rate are usually known.

The clear bore can also be determined by the space requirements of the experiment; 4 cm is about the smallest convenient size. There is in any case little point in making it much smaller, due to the difficulties in cooling the inner turns and still maintaining a useful current density there.

4.1.1. The optimization analysis gives criteria for determining the shape of the most efficient coil, and hence the dimensions. If the magnet is to work at a high-power level, the published results, which assume a constant space factor, are only a rough guide; the optimum current distributions will tend to be less "peaky" and the optimum dimensions to be rather greater. However, the object is rarely to generate the highest possible field for its own sake, but to produce a reliable research instrument—sacrificing, if necessary, some performance to achieve this reliability. By making the magnet larger than optimum, several advantages may accrue.

In the larger magnet the power densities will be reduced. Thus either the cooling is better or the relative volume occupied by cooling voids is less.

By increasing the axial length, an improvement in field homogeneity can be made, as is illustrated for a typical Bitter coil by Figure 4.1. (Note that about the origin the field expansion is $H = H_0 + H_2 z^2 + H_4 z^4 + \cdots$, as discussed in Section 2.3.2.)

The cooling is also improved by increasing the coil length. Usually the increase in cooling area will be more than enough to offset the decrease in water flow which results from the higher hydraulic resistance. The increase in cooling area should also increase the "burnout" limit of the coil (cf. Section 3.2). If further the cooling ducts are fully optimized (Section 3.1.2), those in the longer coil are larger, but fewer in number, and thus easier to manufacture.

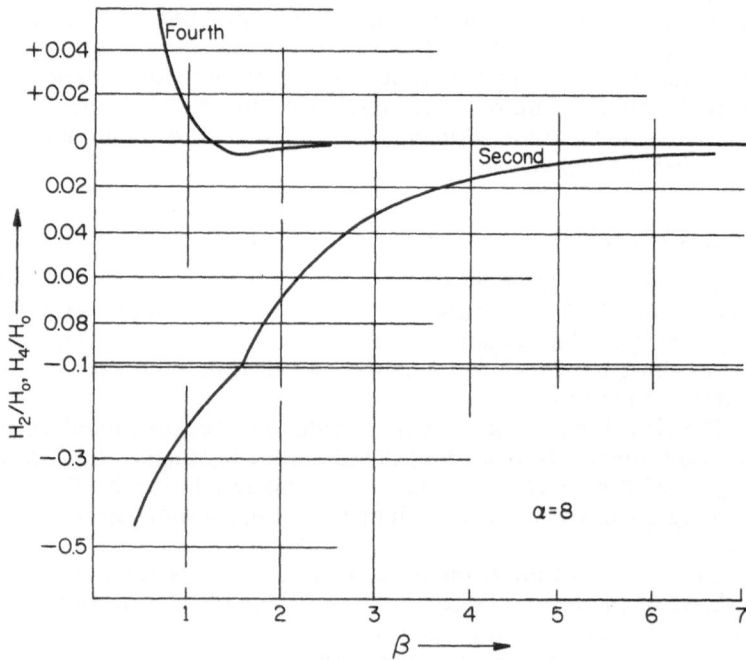

Fig. 4.1. Error coefficients in Bitter solenoid.

Given the cooling fluid, its pressure, and the solenoid dimensions, the results of Section 3.1 can be used to decide the size of the cooling ducts, although in practice (and especially in the case of Bitter-type coils) the optimum may be too small to manufacture conveniently, and some compromise is necessary. The total number of cooling ducts needed is most easily calculated from the coolant flow desired, this in turn being determined by the power to be dissipated and the temperature rises which are considered safe. The ducts will be distributed in such a way as to cool the magnet as uniformly as possible; in practice it is simplest to arrange for all ducts to carry off equal amounts of heat, although this is not quite the same, because it neglects the temperature gradients set up in the conductor metal.

4.1.2. The conductors of normal solenoids are invariably of pure copper, for as the Fabry equation (1.1) shows, the highest possible conductivity is needed. For strength and for ease of fabrication, if punching or machining operations are involved, the metal should be of hard temper in preference to annealed; the resulting loss of conductivity is only a percent or so. The strength to be gained by alloying is not great unless a considerable increase in resistivity can be tolerated. Nevertheless, figures for some possibly useful alloys are given in Table 4.I.

It is probably best to sidestep the strength problem as far as possible by modifying the solenoid geometry, as discussed in Section 2.4.4. One then also alleviates the cooling problems rather than accentuating them. However, some consideration should be given to the use of silver-bearing copper, not for strength, for that is the same as that of pure copper similarly work-hardened, but for its creep-resistance at moderate temperatures. Its annealing temperature is in the region of 300°C, as opposed to about 150° for pure copper; thus the alloy retains its temper at normal working temperatures and can also be quite easily soft-soldered with no significant loss of strength.

It is also possible, in some circumstances, to reinforce the windings by using insulators which are much stronger than copper. Bitter (1962), for example, has proposed using thin stainless steel sheet coated with an insulating resin in place of the more usual insulating fabrics. It appears that this might as much as double the strength of the assembly.

4.2. SOLENOID CONSTRUCTIONS

4.2.1. The obvious way of making a solenoid is to wind copper wire (of rectangular cross section) in layers on a cylindrical former. Adjacent turns are separated by a strip of insulating foil interwound with the copper, and adjacent layers by insulating rods running axially. Coolant flows axially between these spacer rods. In such an

Table 4.I. High-Conductivity Copper Alloys

Alloy	Mechanical properties				Conductivity
	Elastic limit	0.1 % Proof Stress	Ultimate tensile strength	Annealing temperature (°C)	
Copper, tough pitch high conductivity					
O	4	16	56	150	100
H	48	84	100		97
Silver bearing (0.08 % Ag)					
H	48	84	100	300	97
Cadmium copper (0.7 to 1 % Cd)					
H	60	120	140	150	80
Chromium copper* (0.5 to 1 % Cr)	80	110	130	350	80–85
Low beryllium copper* (2 to 3 % Co, 0.35 to 0.7 % Be)	—	220	220	350	45–50
Zirconium copper* (0.15 % Zr)	—	110	120	> 500	90

Strengths are taken as relative to the ultimate strength of hard copper, 25 tons/in.^{2}, 40 kg/mm^2, or 3.9 × 10^9 dynes/cm^2.
O = Annealed
H = Fully work-hardened
* In the fully heat-treated condition.
† After Saarivirta (1963).

arrangement Giauque and Lyon (1960) report that the cooling passages appear to have properties very little different from those with smooth walls. The passage dimensions can easily be made of the order of one or two millimeters, so efficient cooling is possible. Fields in the region

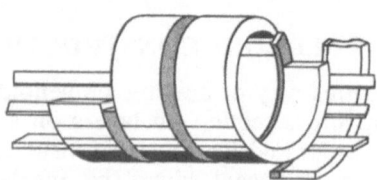

Fig. 4.2. Strip-wound solenoid.

of 100 kOe have been generated in a 10-cm bore. In a later development of this construction, Wood is understood to have generated 120 kOe for only 2 MW dissipation.

4.2.2. Copper strip may alternatively be wound into a spiral, rather as a watchspring. The axial length is made by stacking several such coils side by side. In one such construction, originally due to Tsai and as developed by Wood (1962*a*), the strip is wound with nylon monofilament with such a coarse pitch that it spaces the turns and yet leaves axial cooling passages (Figure 4.3). When several coils are stacked, the passages do not align, but the coolant can nevertheless flow without encountering excessive resistance. Such a construction is not very suitable for very high fields, for under the action of the electro-magnetic forces, the turns tend to move slightly and abrade or crush the nylon. The authors have observed this at fields as low as 80 kOe.

In an alternative form of construction the insulation is a tape—for example, of a film such as "Melinex"—wound with the copper. The coolant flows in slots rolled across the copper strip, as shown in Figure 4.4. The weakest feature of this design is the need to have

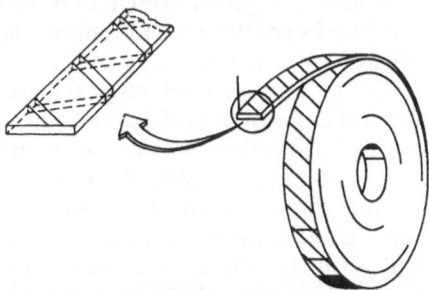

Fig. 4.3. "Tsai" coil.

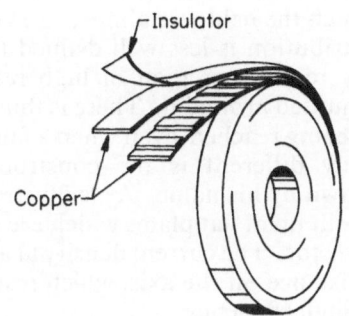

Fig. 4.4. Alternative strip coil.

additional spacers between the coils which are robust and at the same time will allow free coolant flow.

A third construction was used by Purcell and Payne (1963) for an aluminum magnet designed to be cooled by liquid hydrogen. The aluminum foil was wound up with a strip of capacitor paper for insulation, and the whole impregnated with epoxy resin. Radial cooling slots were machined into the coil faces, all burrs being removed by etching in sodium hydroxide.

While it is simplest to use a strip of uniform cross section for the windings, geometries other than the "uniform current density" can quite easily be realized. Strips of different section can be used in different parts of the coil assembly, so that the resulting current distribution is a step function approximating to that desired. As Wood has shown, quite a simple coil need be very little inferior to those usually considered as of higher efficiency. A further property of the strip-wound solenoids is that the current distribution is rather well defined, particularly if a strip of small cross section is used (the resistance can always be made correct by connecting coils in parallel). The associated minor disadvantage is that there is no alternative current route around any high-resistance regions which may develop—e.g., by a local bad contact or blockage of a coolant space—so that such faults are likely to lead to a disastrous failure.

4.2.3. Similar to the strip-wound magnets are the tape-wound coils of Kolm (Bitter 1962), Laquer (1962), and others. Now, however, there is just a single coil, wound of strip of width equal to the magnet length. Space for the coolant is usually made by rolling slots across the strip, or by inserting strips of insulator running axially as the coil is wound up. In Kolm's magnet the copper tape was so tapered in width that the magnet had a trapezoidal cross section. The current in such magnets is then distributed inversely as radius, and the Fabry factor is high. However, to achieve very high efficiency, very high current densities are needed, and in Kolm's case the burnout limit was only about 1.9 MW, at which the field was almost 130 kOe.

The current distribution is less well defined than in strip-wound solenoids; if for any reason a region of high resistivity occurs, the current tends to be shunted around it. There is thus now the possibility of a stable, safe state being reached, even when a fault does arise.

4.2.4. Completely different is the construction originated by Bitter (1936) and known by his name. As is illustrated in Figure 4.5, a conducting helix is built up of flat plates which are overlapped to make contact over a small sector. The current density at any point is inversely proportional to its distance off the axis, which results in the solenoids being of moderately high efficiency.

The plates are separated by insulating disks which have the overlap

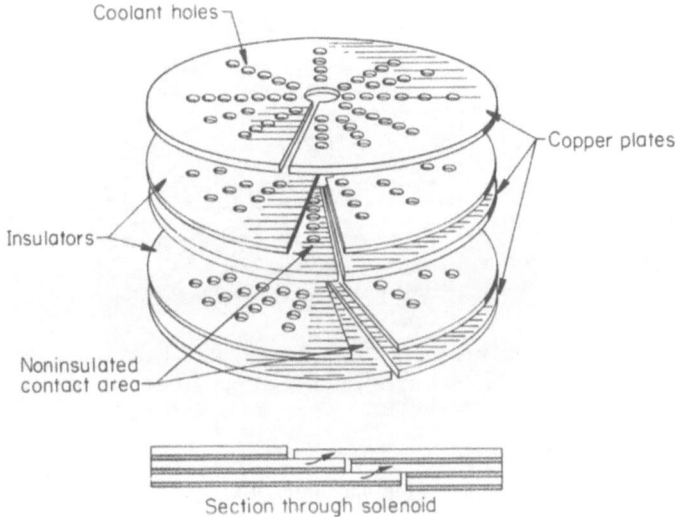

Fig. 4.5. Bitter coil.

sector missing. Suitable insulating materials are those based on fiber-glass, such as 0.1 mm-thick, silicone-varnished glass cloth, a similar material with a layer of mica flakes stuck to it, or a PTFE-impregnated glass cloth.

Pressure contacts are normally used, the stack of disks and insulators being clamped firmly between massive end blocks. In operation the coil pulls itself together, and some permanent contraction occurs as the varnish in the glass cloth "beds in." Elasticity in the tie bolts, or other form of spring loading, is desirable to take this up and ensure that pressure is always maintained. Even so, the contact resistance appears to increase slowly throughout the life of a magnet, apparently as the copper oxidizes. Gold plating of the contact sector has been used (Swim, 1962), and appears to be of some value; it would be even more attractive to form a permanent joint, say by letting suitably plated surfaces cold-weld, but the practical difficulties are considerable.

In the classical Bitter magnets the cooling passages are formed by punching small holes in both conductor and insulator disks, such that when all the disks are stacked, the holes align and the coolant can flow axially. The most favored hole layout today is a honeycomb arrangement, in which each hole is at the center of a roughly hexagonal cell; the cell size and distribution are such that each hole carries off the same quantity of heat. This is almost, but not quite, the same as having the disk at a uniform temperature, for, due to the temperature gradients

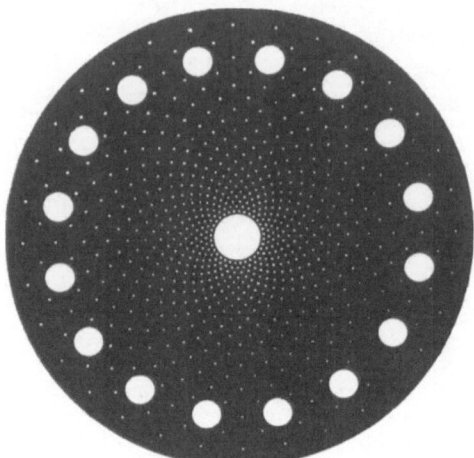

Fig. 4.6. Bitter disk.

in each cell, the mean conductor temperature increases slightly away from the axis (cf. Section 3.3). Such a hole array is seen in the photograph Figure 4.6.

In principle, the Bitter magnet has both a robust structure and an excellent hydraulic configuration. In practice this latter is marred by the roughness of the cooling passages, and friction factors up to ten times those in smooth tubes are found. It is, nevertheless, the type of magnet most used to generate fields above 100 kOe.

Several variants of the Bitter magnet have been made. One, due to Montgomery (Bitter, 1962), employs radial cooling channels. The arguments in favor of this are that such a coil is better able to withstand the electromagnetic forces, that the channels lie along lines of constant potential so that in the event of a blockage shorting of turns is less likely to occur, and that it is better cooled, because the cold water is made to enter at the center where the power density is greatest and thus the cooling most needed. A further advantage is that the slots, whether made by chemical etching (Bitter, 1962) or by radial spacers inserted into a helix machined out of solid copper (Fakan, 1962), should be very smooth and give much less resistance to flow than in the axial flow arrangement. However, it is not easy to feed all the slots and yet neither use too much of the bore nor create such turbulence that some of the slots are starved of water.

Montgomery has also employed a different method of stacking the disks, illustrated in Figure 4.7. Essentially two interleaved helices are made of butt-jointed Bitter disks, the butt joints being so displaced that

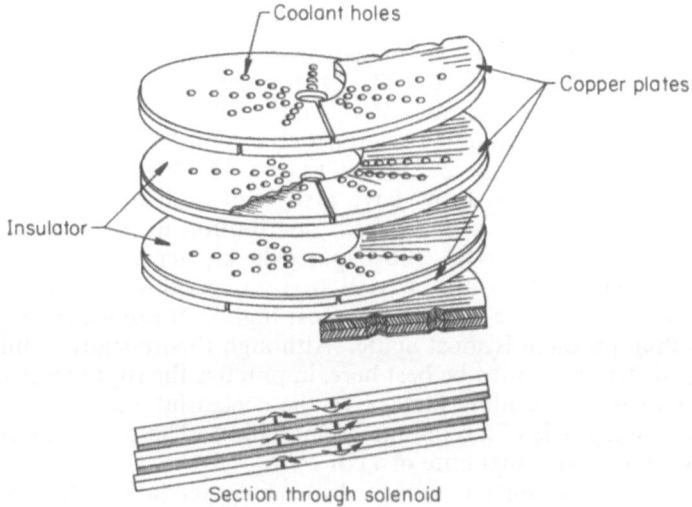

Fig. 4.7. Montgomery variation of Bitter coil.

each is reinforced by the other helix. Although the plates are in contact over their full area, probably only a small fraction is electrically effective, and the overall effect will be little different from the classical arrangement.

A further variant, proposed by Gaume (1962), utilizes plates of different thickness to give an axial variation of current density. Such coils can be highly efficient, such as one at the Naval Research Laboratories in Washington (Swim, 1962), with which 150 kOe was generated for only 3 MW dissipation.

4.2.5. At the highest steady fields there are a number of reasons why a single coil is no longer suitable. We have seen that, as the power and stress levels are raised, it becomes increasingly desirable to modify the current density distribution, which is most easily arranged by dividing the coil into a number of independent sections. One can also then easily incorporate different materials or constructional methods which may be more appropriate to conditions in the various sections. A further point is that the power supply at levels much greater than 3 MW will almost certainly be subdivided into independent units, and even where these can be operated in parallel, single terminals on magnets to carry tens of thousands of amperes are cumbersome. Thus, even though the magnet may be cooled by one water system, the concept of several nested, electrically independent sections occurs quite naturally.

Such a construction is typified by the M.I.T. 220-kOe magnet (Bitter, 1963), which has three stages. The outer, where forces and power densities are low, is a relatively cheap and easily made strip-wound coil—in fact the power density is so low that hollow conductors are used. The designed field is 50 kOe for 4 MW dissipation, in a coil of 90 cm outside diameter and 36 cm bore. The two inner coils are of Bitter construction and in the original design were both radially cooled and stacked in the later manner of Figure 4.7. We understand that axially cooled inner sections have also been built and used. From efficiency considerations the current distribution in the second stage should be of the Bitter type. It dissipates the greater part of the power, being designed to consume 8 MW to produce about 100 kOe in a 16 cm bore. The inner stage is the most highly stressed, and is where the cooling problem is most acute. Although theoretically a uniform current distribution may be best here, in practice the ruggedness of the Bitter coil is the deciding factor. In the successful 220-kOe magnet, this section is made of zirconium–copper alloy. The power dissipated is about 3 MW in a final bore of 4 cm.

The whole magnet is mounted in one water jacket. The M.I.T. arrangement is particularly neat, in that the flow and return pipes enter opposite each other on a horizontal axis perpendicular to the magnet bore; the magnet can be rotated about this axis to give consider-able freedom of access to the working volume. The present authors prefer to separate the water feeds to the various sections, in order to localize the damage should any section fail. It does, however, lead to considerable complication, and makes the provision of adequate mechanical support for the inner coils difficult.

4.3. INSTRUMENTATION

4.3.0. There are two aspects of the instrumentation of a high-power magnet installation which interest us here. The first is concerned with the plant, while the second is the measurement of the magnetic field.

On the plant it is necessary to be able to monitor the cooling circuit parameters such as flow rates, pressures, coolant temperatures, and also voltages and currents for power control purposes. For all these, industrial instruments are adequate and readily available. There is some virtue in having recording instruments, both for helping to reconstruct the sequence of events in any magnet failure, and to make more obvious the slight flow or current irregularities which might give warning of impending failure. In this respect the sound created by the water flow through the magnet can also be very revealing, for the ear is quite sensitive to small changes or to new noises. The beginnings of internal arcing can often be detected by the use of a stethoscope or small microphone mounted against the case of a magnet.

4.3.1. Of more scientific interest is the measurement of the field itself and the corresponding magnet current. In general, the field–current characteristic of a magnet will be measured as it is tested—i.e., it will be calibrated—and in subsequent experiments the stability and perhaps even the linearity of this characteristic will be assumed. Any gross change in this characteristic is indicative of a fault, and any non-linearity is usually due to changes in the current distribution through thermal effects. Thus the current should be measured to better than industrial accuracy.

For measuring heavy currents two techniques will be considered. Either the voltage across a series resistor (shunt) is measured, or the current controls a magnetic amplifier of which the output is readily measured by normal instruments (transductor, or dc current transformer).

If the shunt is of low value, cooling will be no problem, and indeed it may consist of short lengths—about a centimeter—of manganin brazed into a busbar run. The individual elements are connected through a network of resistors designed to average out the effects of maldistribution of the current among the leaves of the bus. A typical high-quality shunt has an output of 10 mV at 16,000 A, with about 0.03 % accuracy.

Higher voltage shunts, such as may be needed in a precision control system, will dissipate such large amounts of power that cooling is necessary. The heat transfer rates involved need not, however, be particularly high. Typical of such shunts is one described by Smith *et al.* (1964). Calibration is a problem, but not if a lower order of accuracy than stability can be tolerated, as will often be the case. A normal procedure would be to calibrate at a low current, determining the temperature dependence also, and extrapolate to higher currents by making any necessary temperature corrections.

The essential circuit of a transductor is shown in Figure 4.8. Magnetic cores with a square hysteresis loop and low coercivity are used. With no direct current flowing in the busbar, the cores can just support the auxiliary ac voltage without saturation. With direct current I_d present, the cores saturate during alternate half cycles, transformer action in the unsaturated core limiting the secondary current to I_d/n. In the ideal case the rectified output is thus the busbar current transformed by the core turns ratio. In practice the cores are not perfect; neither is the auxiliary supply perfectly stable. The useful linear range may be from about 20 to 120% of the designed rating, with the accuracy rapidly becoming very poor at the low end, and complete saturation of the transfer characteristic occurring at the high end. In the useful range an overall accuracy of 0.5% can be readily achieved.

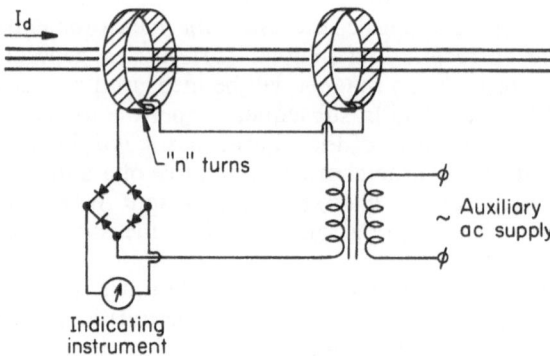

Fig. 4.8. Basic transductor circuit.

For precision control the only magnetic device which can be considered is the second harmonic magnetic modulator described by Williams and Noble (1950); otherwise a high-output (at least 1 V) shunt must be used.

4.3.2. The simplest and most direct methods of measuring the flux in a solenoid are based on integrating the output from a search coil, either electromechanically with a Grassot fluxmeter, with which an accuracy of about 0.5% can be achieved, or electronically. The electronic instrument is more accurate and more convenient in use. An example is the Type J Magnetometer, made by Newport Instruments,* for which some published performance figures are:

Range.................................. 0–160 kG
Drift rate.............................. 1/4 G/min
Accuracy, long-term, with moving coil indicat-
    ing meter.......................... 0.1%
Accuracy, short-term, with digital voltmeter 0.01%

To define the origin for integration the coil can be brought into the magnet from a field-free region, or, where the electronic integrator is used, the coil can be left in the magnet during the field runup. In another method the search coil is rotated at a constant speed in the field, generating an ac signal proportional to this speed and to the field. The signal is picked off with slip rings, and fed to an indicating instrument such as a valve voltmeter. The Rawson fluxmeter, employing this principle, is offered with an accuracy of 1%.

With neither type of instrument is there an intrinsic upper field limit, provided that the search coil voltages do not become so large

---

* Newport Instruments Ltd., Newport Pagnell, Bucks, England.

as to overload the electronics. Where the higher precision is sought, three points need special attention. The first is that the coil must be accurately aligned in the field—an error of $2\frac{1}{2}°$ in direction corresponds to a 0.1% error in the measured field. The second is in the determination of the effective area of the coil; for this the search coil would be used to measure the field of a long, wide-bore, single-layer solenoid of which the field–current characteristic could be accurately calculated. The third arises from inhomogeneity of the field over the volume of the search coil. It is theoretically possible to construct coils which are completely insensitive to this inhomogeneity and given an output proportional to the flux density at the coil center only. This is discussed by Brown and Sweer (1945), and such a coil was made by Williamson (1947). A more practical approach, from which suitable dimensions for a cylindrical coil can be determined, is that of Rezanka (1963) or Pearson (1962).

Search coils are also the most natural devices for monitoring the conditions in a pulsed magnet. Due to the high fields and fast rate of rise, only a few turns on a small former are needed—perhaps only a single turn in extreme cases (cf. Section 8.6.2). Also, because of the short times involved, the integrator can be a simple resistance–capacitance network, with the output displayed on an oscillograph and photographed. A good example of such a system is described by Furth and Waniek (1956).

4.3.3. For the most precise measurements of magnetic field, one must rely on some form of magnetic spin resonance technique.[*] For electron spin resonance, taking the Landé splitting factor g as 2, the relation between resonance frequency and field is

$$f(\text{Mc/sec}) = 2.80H\,(\text{Oe}) \tag{4.1}$$

For proton resonance the corresponding relation is

$$f(\text{kc/sec}) = 4.26H\,(\text{Oe}) \tag{4.2}$$

We see that for fields in excess of 50 kOe, electron spin resonance frequencies will be greater than 140 kMc/sec, already inconveniently high for ordinary laboratory practice. On the other hand, for proton resonances in the field range 50–250 kOe the frequency range is the much more favorable one of 0.2 to 1.1 kMc/sec. If another nucleus, such as that of lithium, with a spin of $\frac{3}{2}$ is used, these frequencies may be reduced by a factor of 3.

The very great precision of the technique can lead to difficulties, for only a very small field variation over the specimen volume can be

[*] For a general description of these techniques, see Ingram (1955).

tolerated. At best the broadening of the resonance leads to loss of accuracy; at worst no resonance at all is seen. Hitherto very little work has been reported where such precision at fields in the 100 kOe region has been required or attained.

4.3.4. The Hall effect provides the basis of a very useful method of monitoring a magnetic field. The sensitive element is usually a thin rectangular plate of a semiconductor. Originally germanium was used, but now a mixed indium arsenide-phosphide is found to give better sensitivity and temperature stability. In Figure 4.9 the control current I_c, the field H, and the Hall voltage V_H, are all mutually perpendicular. To a first approximation

$$V_H = \alpha I_c H \tag{4.3}$$

where α is a constant depending on the material properties and plate dimensions. In practice the device is nonlinear, though by proper choice of dimensions and load resistor, R_H, this can be corrected at low fields (Kuhrt, 1954). A typical output may be 0.01 V/kOe, with temperature coefficient of 10^{-3}/deg C.

The advantages of the Hall probe are its compactness and that the output signal is proportional to the field, without the need for further processing, over a frequency range limited by the external circuit.

4.3.5. Many other phenomena are known which show a magnetic field dependence and could thus be used as the basis of a field monitor.

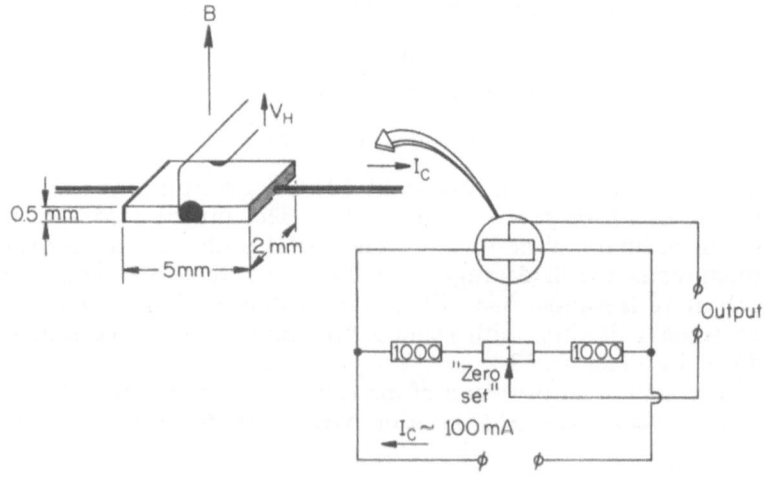

Fig. 4.9. Hall probe and circuit.

Although none are widely established, we may mention two which have found some application. The first is the magnetoresistance effect; bismuth can be used to make very small sensitive probes (Keller, 1953), but the temperature dependence is great. For higher fields, copper has been used.

Another effect which has been investigated into the megaoersted region is the Faraday rotation in glass and quartz (cf. Section 8.6.4). As a field monitor it has the advantage that no leads to a probe are necessary, and by using the very fine parallel pencils of polarized light one can obtain from gas lasers, the specimen can be made extremely small.

Chapter 5

Power Supplies

5.0. INTRODUCTION

In setting up a magnet installation considerable engineering effort is needed for both the power supply and the cooling plant. For a 100-kOe facility we have seen that the power required is around 2 MW, and for 250 kOe around 12 MW, which is within the realm of heavy industrial equipment. Yet for some experiments the stability demanded may be that of precision laboratory equipment, being parts in 10^5 or even better for periods of a few minutes.

To reach still higher fields by conventional means, even higher powers are needed. Fortunately the fundamental limit to power ratings is usually set by thermal behavior, and most heavy plant has sufficient thermal inertia to permit gross overloading for short periods. A four times overload for 20 sec (Adams, 1962) may be permissible after only small departures from standard industrial practice, and represents a very useful extension to the capabilities of a system for a very modest increase in cost.

5.0.1. In the first section cooling systems and power supplies for conventional solenoids will be discussed. In the second section cryogenic magnet installations will be considered. The power supplies are similar to those for conventional solenoids, but the reduced power rating can allow a wider range of stabilization techniques to be practicable. The cooling plant, on the other hand, now becomes the major installation where continuous working is desired.

Superconducting solenoids, in principle at least, need no power supply to maintain the field. However, the field must be set up, either by flux-pumping (Section 7.6) or by exciting the coil from a current source. A battery and series resistor are sufficient, but it is more usual to have an electronically stabilized device, with circuits for the protection of both coil and power supply should the superconductor be accidentally driven normal.

The final choice between systems will often rest on cost, but this varies so greatly with local conditions and requirements that no general conclusions can be drawn. Here we can only outline the factors involved.

5.1. CONVENTIONAL COILS

5.1.1. For conventional solenoids, working at near ambient temperature, the most widely used cooling medium is water. It is cheap, readily obtainable, inert, and nontoxic, and has the further virtue of being one of the most efficient coolants known (cf. Section 3.4).

Power densities and heat fluxes are so high that all unnecessary thermal resistances must be eliminated, and in practice the coolant flows directly over the current-carrying conductors. Thus it is usual to treat the water in an industrial de-ionizing plant until its resistivity is 10^6 Ω-cm or higher. Even so, some electrolytic action can take place, and it is considered prudent to keep the supply voltage below about 250 V. However, in at least two laboratories in which low-voltage supplies are used,* the coolant is taken straight from the city water supply, with apparently no serious corrosion troubles.

The avoidance of corrosion seems to be something of a black art, because experience differs widely. The general rule is to avoid using metals of widely differing potentials; we have observed severe corrosion of aluminum parts inadvertently introduced into plastic and copper system, and we now use plastic or plastic lined pipework, with aluminum bronze or stainless steel for parts such as pump housings and impellers. It also seems to be good practice to design magnets so that pockets of stagnant water cannot form. Nevertheless, we understand other laboratories (e.g., Wood, 1962a) to have no trouble using iron or aluminum alloy pipework. Deoxygenation of the water is probably more trouble than it is worth, for one cannot then use stainless steel for fear of rusting, and the cooling system must be kept sealed.

Where city water is used, the mains pressure has proved adequate. Generally, however, a pump delivering the required volume (say 400 gal/min or 0.03 m³/sec for a 2 MW magnet) at a pressure up to about 100 psi (or 7 atm) is needed. This should be of the centrifugal type, to minimize as far as possible undesirable noise and vibration. For the same reason, care should be exercised in the layout and shaping of components in the hydraulic system to avoid unnecessary turbulence and cavitation, such as using swept T's and avoiding sharp bends. Other components in the hydraulic circuit include a heat exchanger, the secondary of which is cooled by raw mains or river water, or by a

---

* The Mond Laboratory, Cambridge, with 80 V and NASA Lewis Research Center, Cleveland, with 30 V.

second circuit which includes a large pond or an evaporative cooler, and storage tanks for the demineralized water, which also usefully increase the thermal capacity of the system.

In two laboratories, however, corrosion problems are avoided by using alternative coolants. At the University of California, Berkeley, Giauque (Giauque and Lyon, 1960) has a motor generator delivering 700 V and has chosen kerosene as the coolant, while at the Kamerlingh Onnes Laboratory in Leiden (de Klerk, 1962) the power source is a 1000 V mercury-arc rectifier, and Dowtherm, a commercial coolant based on orthodichlorbenzene, is used. Both are less efficient heat transfer media than water; in addition, kerosene presents a fire hazard, while Dowtherm is expensive and its vapor is evil-smelling and toxic.

5.1.2. Most magnet power supplies use voltages in the region of 200 V. Lower voltages imply very high currents, and thus require expensive and unwieldy busbars and switchgear, particularly if the current is to be made available at several stations.

We shall think of all power supplies as being from the public supply—the alternative of generating one's own power, using, for example, gas turbines or diesel engines as prime movers, is not economically attractive unless the laboratory is situated a considerable distance from a suitable supply point, and even then one forgoes the very high short-term overloads available on a grid with a large connected generating capacity. This is probably true even of daytime working, though electricity tariffs are heavily weighted to favor off-peak hours.

For conversion of the high-voltage (typically at least 11 kV) alternating current to low-voltage direct current, many types of industrial plant of suitable rating are manufactured. They may be divided into rotating machines—i.e., an ac motor driving a generator; static machines, such as step-down transformers followed by rectifiers; or storage devices, such as lead–acid cells.

5.1.3. New motor generators will usually be offered with synchronous driving motors, these being rather more expensive but more efficient and capable of running at a better power factor than induction motors. For a little-used installation the savings due to higher efficiency would not be sufficient to outweigh the difference in initial cost. If overloads of several times full load are to be drawn, the machine must be suitably designed, will be somewhat larger than a standard machine, and cost perhaps 10% more. Adams (1962) has pointed out that when a large well-interconnected power network exists, as in the U.K., these overloads can be taken up by the inertia of the system. Obviously the cooperation of the generating authority is required! The alternative approach, that of using massive flywheels, is only to be recommended where the electricity supply system is incapable of absorbing the power variation without appreciable interference with other consumers—as, for example, at the American National Magnet Laboratory in Cambridge, Mass. There·the pulses demanded are up to 32 MW, but the local generating capacity was only 100 MW at the time of installation. The inertia of the rotating parts alone is sufficient to reduce the effects on the output of any mains voltage fluctuations to very small proportions, and in this respect synchronous machines are probably slightly superior to induction motors.

The costs quoted for rotating machinery will usually include exciter gear and erection and commissioning, but will exclude the costs of buildings and foundations—typically an additional £10,000 for a 3-MW set if starting from scratch. Some allowance should also be made under running costs for regular light maintenance, and occasional major work if the machine is in full-time use.

The generators most commonly used are conventional commutator machines, but homopolar dynamos and alternators followed by semiconductor rectifiers have also been proposed. Commutator machines for the ratings required are large, low-speed (300–500 rpm) devices, becoming larger and slower as the rated voltage is lowered and the rated power increased. Rebuilt second-hand machines can often be obtained at prices far less than new, but cannot be expected to be suitable where the highest quality output is needed. The output spectrum of such a machine is quite complex, as may be seen from Figure 5.1, which shows a typical spectrum for generators of good quality. The ripple components arise at harmonics of four principal frequencies: the rotational frequency of the machine and the pole, slot, and commutator segment passing frequencies. The first two arise from slight electrical eccentricity and asymmetry, but with careful design and construction can probably be reduced to a few parts in 10,000 of the open circuit voltage, as in the M.I.T. machines. The third component can be greatly reduced by skewing the slots, and is not seen at all in the M.I.T. machines. The commutator ripple is of widely varying amplitude, is worse with an inductive load, and can be kept small only by careful trimming and maintenance. The commutator should in any case be sized for the maximum overload current and not the normal rated current.

Control of the output is usually by control of the excitation, the power needed being of the order of 40 kW for a 2-MW generator. The time constant is long, due to the high inductance of the field winding However, by using the technique of "field forcing," a faster rise time can be achieved, but only at the expense of increasing the voltage rating of the exciter. There is in any case a limit of a few tenths of a second set by eddy currents in the machine frame, though this

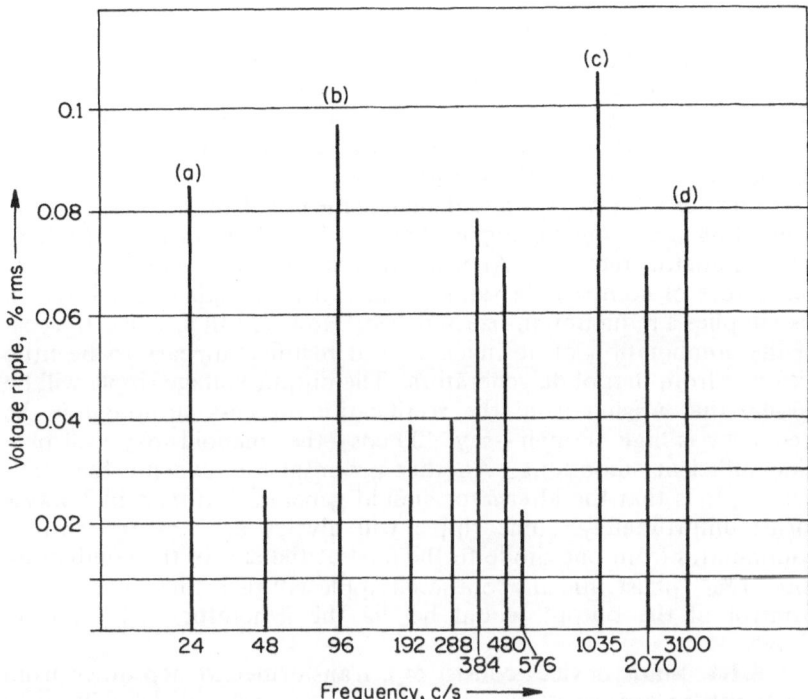

Fig. 5.1. Voltage ripple of generator on inductive load. (a) Rotational frequency;
(b) pole passing frequency; (c) slot frequency; (d) commutator segment frequency.

can be reduced a little by going to the more expensive laminated
frame construction. Standard excitation equipment of the amplidyne
type also has a long time constant, and in a closed loop control system
the overall figure will, therefore, be quite large. However, modern
excitation equipment using controlled semiconductor rectifiers might
offer an improvement in this respect.

In the event of a short circuit on the dc side, the current can build
up very rapidly to values many times the rating, and protection by
means of high-speed circuit breakers is essential.

The homopolar generator produces a very smooth output, for
the current is collected by slip rings rather than a commutator; furthermore, the rotational ripple is small and is at a higher frequency than
in conventional machines because a much higher speed of rotation is
usual. Unfortunately, even at the highest practicable speeds, the output
voltage is limited to about 60, or possibly 120 if duplex machines are
used (Klaudy, 1962, 1963), and the busbars needed for high-power

working are thus of enormous* cross section. In theory, more stages could be used in series to give any desired voltage, but at the time of writing these had not been developed to being a commercial reality. In principle, homopolar machines are reliable and robust, but in practice, experience of them seems to be limited and variable. Very satisfactory performance is reported by Fakan (1962).

Further solutions proposed include the use of alternators, possibly generating at a frequency higher than 50 cps, followed by a multiphase semiconductor rectifier. Alternators tend to be cheaper than dc generators of comparable ratings and run at a higher speed, which also implies a reduction in size and cost. However, in practice, the cost of the combination of ac machine and rectifier appears to be little different from that of dc generators. The output voltage ripple will be largely that arising from the rectification process, although if the frequency is high enough—say, 400 cps—the solenoid may well provide sufficient smoothing. Another proposal for reducing the rectifier ripple is that the alternator should generate a trapezoidal waveform; unfortunately, some ripple will always arise as the current commutates from one diode to the next at the end of the conduction time of each phase, and any rotational ripple will also remain unaffected. Control of the output would be via the generator excitation, as above.

5.1.4. Static devices consist of a transformer to step down from the supply voltage to a low voltage feeding a rectifier, which will today employ silicon junction cells. Often a voltage-regulating transformer is included, on the high-voltage side.

In the earlier installations of this type, mercury-arc rectifiers are used. Consequently the output voltage is quite high: 1000 V at Leiden (de Klerk, 1962) and 400 V at Tohoku (Maeda, 1962). The output voltage is regulated by controlling the firing time of the rectifiers, so that the output waveform has a very high ripple content, and at both laboratories inductance–capacitance filters are used to smooth it. With additional smoothing from a 3-mm-thick, liquid-nitrogen-cooled copper shield, the short-term stability at Leiden can be better than 1 part in 10^5.

Modern rectifiers employ silicon diodes in place of the mercury-arc valves, and would be designed for a lower output voltage such as 200 V. The preferred industrial circuit is essentially that of Figure 5.2; 12-pulse operation results, giving a fairly smooth output and acceptably low harmonic feedback into the supply network; the combination of star and delta secondaries and bridge rectifier allows the most efficient use to be made of the transformer.

---

* For example one leaf of $6 \times \frac{1}{4}$ in. copper per 1500 A.

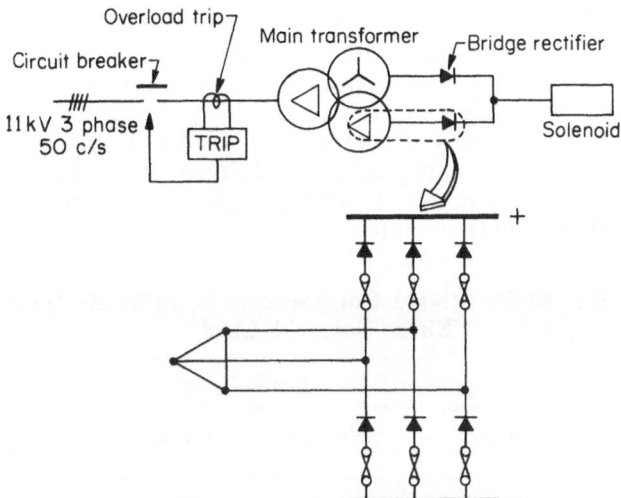

Fig. 5.2. High-power bridge rectifier circuit.

Control is not simple, although at the time of writing, development is proceeding fast. The primary voltage regulating transformer, of which several types are manufactured, is a slow device, typically requiring one or two minutes for full-range operation, although a few tens of seconds should be quite feasible. For full-range use, certain types suffer from the disadvantage of having very high reactance over part of the range. For faster control silicon controlled rectifiers are already made to handle several megawatts at several thousand amperes, just as the earlier mercury-arc valves. However, it will usually be cheaper, and will certainly give smoother output and greater efficiency, if the controlled rectifier is used in combination with a voltage-regulating transformer.

The mains fluctuations and rectifier ripple will remain. Little is published about the former, and in any case it would be of limited value because conditions can vary so widely, both from place to place and even from time to time at a particular location. In our experience, most of the fluctuations are comparable in magnitude to the rectifier ripple (12-pulse), but irregular steps of a percent or so and occasional much larger transients occur. The energy in these transients is quite small, because small filters suppress them to the level where they cannot damage the diodes. The fundamental ripple frequency in a p phase rectifier working from a supply of frequency f is pf, and the fractional rms ripple voltage is $\sqrt{2}/(p^2 - 1)$. However, due to unbalance in either the supply or in the equipment itself, lower harmonics of f can appear. The magnitude of the ripple is in any case greater than the expression quoted suggests, for the current cannot commutate instantaneously from one diode to the next, and some waveform distortion results, as sketched in Figure 5.3. The performance of an actual rectifier with a solenoid load is shown in Table 5.I. The use of controlled rectifiers or some types of magnetic amplifier for regulation will lead to an increase in the higher-frequency ripple components, although in principle at least all those at frequencies below *pf* can be attenuated.

Table 5.I. Major Ripple Components: 12-Pulse Rectifier with Bitter Solenoid Load

Frequency c/sec	Ripple, % rms	
	Voltage	Current
50*	0.12	0.01
100*	0.64	0.035
300*	2.5 to 0.5	0.066 to 0.013
600	1.7	0.033
900	1.4	0.022
1200	0.48	0.007

* Dependent on supply balance.

Saturable reactor control has been proposed in an arrangement in which the control current is well smoothed (see, for example, Milnes, 1957) and thus the output current is largely free of the usual rectifier ripple and is insensitive to supply fluctuations. However, typical component ratings are greater than the power to be controlled

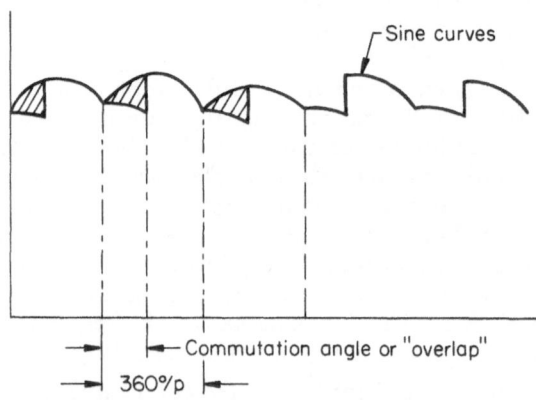

Fig. 5.3. Waveform distortion in rectifier.

if anything approaching full range control is demanded, and a full-scale system would thus be extremely expensive.

5.1.5. From the foregoing (summarized in Table 5.II) it will be clear that for many purposes, rotating machines and rectifier-transformer sets can meet similar specifications, but that on grounds of costs and operating convenience the rectifiers are much to be preferred. The industrial closed-loop control systems could have a stability approaching 1 part in 10^4, but only with a very limited response time. As the stability requirements are made more severe, both in magnitude and in frequency range, the choice becomes less clear-cut, for simple rectifier equipments suffer from two disadvantages. The first is that they cannot attenuate supply voltage fluctuations, and the second, that the current ripple in a typical solenoid is somewhat greater than that from a well-designed motor generator, and is, moreover, at a considerable higher frequency. In theory, it is possible to overcome these disadvantages, but only by incorporating additional equipment which would be costly and would need considerable development effort. The result would be for the rectifier to lose most, if not all, of its cost advantage. Where the very highest stability is essential, the picture is still more confused, for now the output from a conventional motor generator is also inadequate and further smoothing is needed. One possible solution is to pass the whole output through a bank of power transistors (Smith *et al.*, 1964), but it may also now be economic to develop other proposals, such as the homopolar machines.

5.1.6. When the high-field facility at R.R.E. was established, the capacity of the electricity supply was insufficient for any of the power supplies already mentioned to be considered. The solution adopted was to use lead–acid storage cells of several thousand ampere-hour capacity which were available (Parkinson, 1962). Discharge currents of many thousands of amperes may be taken, the limiting factor at present being the switchgear ratings. The output is free of ripple, though there is a slow fall in voltage as the cells discharge. In operation, the current is brought to its final value in a series of steps, by switching resistors in series with the solenoid. This, and adjusting the number of cells in series before switch-on, are the only practicable means of controlling the current.

Given suitable dc circuit breakers, the cells should be capable of supplying extremely high powers for a short time, limited only by their own internal impedance. In such applications so far reported, the cells have been rather conservatively rated (Skellet, 1962).

5.2. CRYOGENIC SYSTEMS

5.2.0. In this section we shall be concerned with cryogenic solenoids—that is, solenoids which are operated at temperatures

Table 5.II. Summary of 3-MW Power Supply Comparison

	ac motor driving			Regulator—transformer—12-pulse silicon rectifier
	Homopolar generator	dc commutator generator	Alternator with rectifier	
Efficiency at full load	90%	88%	84%	96%
no load loss	2%	2%	2%	93% at ¼ full load
power factor	Almost 1, if synchronous motor used ——→			0.96, worse if controlled rectifier used
Overload capacity: ½ hr	←——— Design for 15% voltage and current ———→			
short-term	Very high, dependent on motor	4 × full load current for 10 to 20 seconds	4 × full load current ——→	4 × full load current for 5 sec
Noise and vibration	←——— Usual from rotating machinery ———→			Air cooling noise, negligible vibration
Minimum sweep time	~2 sec	~0.5 sec	~1 sec	~30 sec
Voltage ripple, main components	0.02% at 50 c/sec (rotational frequency)	0.02% at 6 c/sec	2% at 600 c/sec	0.5% 100 c/sec, 0.5% 300 c/sec, 2% 600 c/sec
Response to mains	←——— Smoothed by inertia of rotating parts ———→			No smoothing
Cost: Capital, assuming overload as above, £	40,000	55,000	50,000	35,000
Additional protection gear	←——— High-speed circuit breakers ———→ 4,000	4,000	—	Not needed
	(possibly not necessary)			
Building and foundations	10,000	10,000	10,000	1,000
Maintenance	←——— Routine light maintenance trimming of brush gear and commutator grinding ———→			Almost nil
Control and stability (extra costs)	←——— For £10,000 could meet 1 in 10⁴ specification ———→			For £10,000 could meet 1 in 10³ specification, with slow response to mains steps
	←——— Costs to go to 1 in 10⁵ possibly £20,000 ———→			For £60,000 could meet 1 in 10⁴ or 1 in 10⁵ specification

considerably below ambient. The two coolants most used are liquid nitrogen and liquid hydrogen, although liquid neon has also been used, and compressed helium gas at about 10°K has been proposed (Kronauer, 1962).

5.2.1. As the temperature is lowered the resistivity of pure metal falls, the variation being reasonably accurately described over a wide temperature range by the Bloch–Gruneisen law if the value of the Debye temperature, Θ, is suitably chosen:

$$\rho(T) \propto x^{-5} \int_0^x \frac{s^5 \, ds}{(e^s - 1)(1 - e^{-s})} \tag{5.1}$$

where $x = \Theta/T$. The theoretical resistivities of some metals of interest are tabulated. However, at the lowest temperatures two further effects become dominant, and to a rather rough approximation the resistivity becomes the sum of three independent terms. Apart from that already mentioned there is a residual resistivity, more or less temperature-independent and determined by the impurity of the metal or lack of perfection in its crystal structure, and a magnetoresistance. This is usually shown by plotting $\Delta\rho/\rho_0$ against $H\rho_\Theta/\rho_0$. Here ρ_Θ is the resistivity at the Debye temperature; and ρ_0 is that in zero field at the measuring temperature. Measurements, many in pulsed fields (Lüthi, 1960), have been made on a wide range of metals; for copper the resistivity appears to increase steadily with temperature, but for some metals, notably sodium and aluminum, it appears to saturate at a value several times the zero field value.

Table 5.III. Theoretical Resistivities of Some Metals

Temperature, °K	Resistivity, $\mu\Omega$-cm			
	Silver	Copper	Aluminum	Sodium
373	2.11	2.25	3.77	
293				5.11
273	1.52	1.59	2.61	
81	0.316	0.226	0.293	1.05
20	3.6×10^{-3}	8.15×10^{-4}	7.03×10^{-4}	1.61×10^{-2}
10	1.16×10^{-4}	2.55×10^{-5}	2.20×10^{-5}	5.35×10^{-4}
Debye temperature,[*] °K	223	333	395	202

[*] After MacDonald (1956).

5.2.2. Since the resistivity of copper at 77°K is about one-seventh that at room temperature, and aluminum has been reported (Purcell and Jacobs, 1963) with a residual resistivity as low as 2×10^{-9} Ω-cm, the power requirements of cryogenic coils are clearly far less than those of water-cooled coils. Whether a net saving results depends on the refrigerator power needed to produce the cryogenic environment. Consider a refrigerator which produces coolant at a temperature T_c, has a heat sink at temperature T_h, and works with an efficiency η relative to one employing a perfect Carnot cycle. If this is continuously cooling a magnet dissipating power W, the refrigerator power needed is

$$W_r = W \frac{T_h - T_c}{\eta T_c}$$

The total power requirement is thus $W + W_r$, which is

$$W_c = W \left(1 + \frac{T_h - T_c}{\eta T_c}\right)$$

W_c is to be compared with the power dissipated in a conventional magnet, W_h, which is given roughly by

$$W_h = W_c \frac{\rho_h}{\rho_c}$$

Post and Taylor (1959) show that there are conductor materials for which W_c can be less than W_h over a certain temperature range, although the precise figures depend rather critically on the refrigerator efficiency which can be achieved in practice. They prefer sodium conductors cooled to 10°K, but suggest aluminum cooled to 20°K as another possibility.

5.2.3. In practice, the situation is unlikely to be quite as attractive, for a variety of reasons. First, unless the conductors of the cryogenic coil are of much smaller cross section than those of the water-cooled coil, the current requirements of each are the same. This is not desirable, for it is much easier to stabilize lower current power supplies and to feed the lower currents to the coil without introducing a large heat leak; but if the coil is redesigned to incorporate thinner conductors, either the space factor becomes worse or the insulation must be made much thinner. This latter course was adopted by Purcell (1963) in a rather novel construction.

The fact that the conductors are of pure metal implies that they have little mechanical strength, and it is not known how much strain can be tolerated before the resistivity increases significantly. In the case of sodium it is thought to be large, but the conductors then have no intrinsic strength. In any high-field cryogenic coil there will thus be a significant volume of structural material, say 20% at least if high-tensile steels are to provide all the strength, again lowering the space factor.

If pressure contacts are used anywhere in the magnet, the contact resistance becomes more significant at lower temperatures.

Finally, some consideration should be paid to the precise operating temperatures of the various parts of the system. If we assume cooling to be by a liquid in turbulent flow, the formulae of Chapter 3 may be applied to show that the conductors may be 5 to 10 degrees hotter than the liquid nitrogen coolant, or 1 or 2 degrees hotter than liquid hydrogen. Such experimental evidence as there is seems to confirm this.

5.3. ECONOMIC CONSIDERATIONS

5.3.0. The discussion of cryogenic systems leads naturally into a discussion of the economies of generating magnetic fields. Because the situation is changing fast, largely due to the way the technology of superconductors is advancing, only a few general remarks will be made.

5.3.1. The cost of a magnet installation is most simply broken down into a capital charge and a direct running cost. For comparison with running costs the capital will be written off over a number of years —say 10 or 15. Obviously the main factor in the capital cost is the performance demanded. Thus, up to a certain maximum field, in the region of 100 kOe, the cost of a superconducting coil will be less than that of a water-cooled solenoid, and because the running costs are also so much less, the economic choice is clear.

5.3.2. Now that superconducting solenoids have become accepted as the "regular" laboratory means for generating fields below 100 kOe, there is a tendency to think that it is a fairly straightforward matter of solenoid design to push to near the upper critical field of niobium-tin. However, Montgomery (1966) published data on the theoretical capital costs of building solenoids for fields approaching 200 kOe based on the short sample performance of one of the best Nb_3Sn multistrand cables available using a 3-cm-bore solenoid as a basis. As shown in Figure 5.4, costs rise very steeply at about 175 kOe. At the time of writing, the highest fields so far generated are about 130 kOe. However, the point at which the rise takes place is sensitive to the current density, and the figure shows that if material can be produced with a markedly improved current density, what may be an "economic barrier" can be pushed to higher fields.

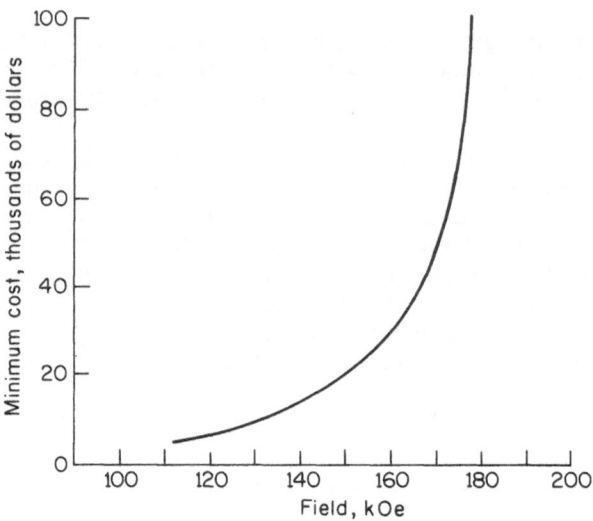

Fig. 5.4. Minimum costs of niobium-tin solenoids (3 cm bore).

5.3.3. As one goes well above 100 kOe, a number of choices present themselves, but superconductors alone are no longer sufficient. Therefore, the cost of a high-power installation has to be faced, with the choice between cryogenic and conventional systems.

For a much-used facility it seems most unlikely that the running costs of a cryogenic system could be less than those of a conventional system, while the capital cost may well be much greater. For a little-used facility a second possibility exists, that of using the cryogenic fluid as an energy storage medium. In the extreme case most of the capital cost would lie in the cryogenic handling and storage facilities, and in the magnets, leaving the savings on electrical and liquefaction plant to offset the rather high cost of expended energy bought as liquid nitrogen or hydrogen.

Although the energy costs are high, so also are those in a little-used conventional facility, due to electricity tariffs which penalize a poor load factor. For example, the present industrial tariff in Britain is about 1d per kilowatt-hour consumed, plus a charge as high as £1 per month per kilowatt of maximum power demand as recorded on a meter averaging over half-hour periods. Even if one chooses to work in "off-peak" hours, some maximum demand charge or revenue guarantee may still be imposed. The precise crossover point is very sensitive to local conditions such as these, costs of cryogenic equipment —always higher than the equivalent in conventional equipment—

and even in local expertise. Possibly the real value of the cryogenic technique is in pulse work where the pulse length—tens of seconds or more—is such that the magnet operates under steady-state conditions. The effective powers which could be realized may be much more than the short-term ratings of generators would permit, or, indeed, more than the supply authority would allow.

Although only a few years ago one would have thought in terms of conventional systems alone for generating the highest fields; now interest is being shown in the possibilities of hybrid systems which include superconducting components. A large-bore superconducting magnet may contain a normal inner section, thus extending the field capability of the existing power supply by perhaps 50 or 100 kOe; or alternatively the normal coil may contain a superconducting insert. In either case the economics are again particularly sensitive to super-conductor costs and to the current densities which either type of conductor can sustain.

Indeed, in a discussion of this question Montgomery (1966) shows that if the superconducting insert is too small the power needed to generate a given field in a given bore may be greater than if no insert were used, although by proper design the hybrid system of either con-figuration can show a power saving over a very wide field range. Nevertheless, the cost of superconductor is not small, and for some magnets may well exceed the extra cost of a bigger power supply.

In Section 7.7.0 recent work shows a possible method by which current densities could be markedly increased.

Chapter 6

Hard Superconductors

6.0. INTRODUCTION

When superconductivity was first discovered, there was an immediate interest in the possibility of generating powerful magnetic fields
using superconducting solenoids and thus in so doing without power
dissipation. This idea was quickly shelved when it was discovered
that fields less than about 1000 Oe were sufficient to destroy superconductivity in the then known superconductors. The first materials
studied were naturally the pure elements such as tin and lead. Table
6.1 shows the currently known superconducting elements with their
transition temperatures and their critical magnetic fields extrapolated
to 0°K. The idea was again expressed when de Haas and Voogd (1929,
1931) showed that alloys of lead and bismuth remained superconducting
in fields near 20 kOe. However, at such fields it was thought that the
current-carrying capacity of the material was too low to be useful.
It was not until Yntema (1955) and Autler (1960) used niobium coils
on iron formers that it was shown that usefully high-current densities
in the presence of magnetic fields could be achieved. Then followed
very rapid developments with the discovery that the compound
niobium–tin, Nb_3Sn, could carry very high "supercurrent" densities
($\sim 10^5$ A/cm^2) in fields near 90 kOe and that alloys such as those of the
niobium–zirconium system also had extremely useful behavior
(Kunzler, 1961).

The useful feature which all such materials have is a very pronounced shoulder in the critical current–magnetic field curve (Figure
6.1). At the shoulder the critical current density is sufficiently high to
permit solenoids to be built which can generate correspondingly high
fields. In parallel with these practical discoveries theoretical developments have taken place which enable them to be understood qualita-
tively in relation to the more familiar properties of the pure elements
like tin and lead. The sharp transitions of such elements are now

77

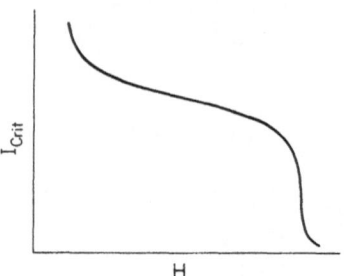

Fig. 6.1. Critical current *vs.* field curve
for hard superconductors.

Table 6.I. The Superconducting Elements

Element	T_c °K ($H = 0$)	H_0 (oersted)
Aluminum	1.19	99
Cadmium	0.56	30
Gallium	1.09	51
Indium	3.41	283
Iridium	0.14	
Lanthanum α	~5	
Lanthanum β	~5.95	1000
Lead	7.19	803
Molybdenum	0.96	
Mercury α	4.15	411
Mercury β	3.95	340
Niobium	9.46	1944
Osmium	0.7	1070
Rhenium	1.70	201
Ruthenium	0.49	66
Tantalum	4.48	830
Technetium	11.2	
Thallium	1.37	162
Tin	3.72	306
Titanium	0.40	100
Tungsten	0.001 to 0.01	
Uranium α	0.6	~2000
Uranium β	1.80	
Vanadium	5.30	1310
Zinc	0.88	53
Zirconium	0.75	47

regarded as characteristic of type I or soft superconductors, as opposed
to type II; these terms will be explained more fully in a later section.
Present thoughts are that type II superconductors in the hard worked
state are those which are useful for use in solenoids.

In the following section the salient properties of superconductors are outlined to provide the basis of a discussion of materials which can be used in "supermagnets" (superconducting solenoids), and some of their limitations. The difficulties of constructing useful supermagnets are dealt with in the next chapter.

6.1. SUPERCONDUCTIVITY

6.1.0. Besides 24 elements which are known to be superconducting, there are not less than 400 alloys and compounds which also show this property, some of them even made from elements which themselves are not superconductors, e.g., Au_3Bi. The first characteristic of a soft superconductor is that the resistance falls to zero at a well-defined critical temperature (Figure 6.2a), which is, however, a function of magnetic field (Figure 6.2b); T_c is the value in zero field). The curve relating H_c, the critical field, with temperature T, is to a good approximation parabolic with its maximum value H_0 at $T = 0$. Values of H_0 for the elements are given in Table 6.I. At the superconducting transition there is a specific heat anomaly or discontinuity (Figure 6.2c), below which the electronic specific heat follows an exponential law. However, the distinguishing characteristic of a soft superconductor is the exhibition of the Meissner effect, which sets it apart from a perfect conductor. If a perfect conductor in a simple geometrical shape is cooled from a point such as B in Figure 6.2b to a point such as C (without varying the magnetic field), the flux within the material remains unchanged. On removing the external field, surface currents flow in the material which maintain the flux in the body constant. With a soft superconductor on passing from B to C, at the transition curve X, the flux is expelled from the body, and for all points lying within the transition curve, $B = 0$ inside the body. This is the Meissner effect. Superconducting surface currents flow in the body to a degree just sufficient to enable this to happen. On removing the field, the surface

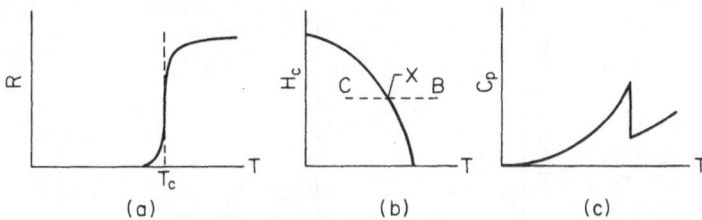

Fig. 6.2. (a) Critical temperature of a soft superconductor shown by the resistance *vs.* temperature curve. (b) Critical field of a soft superconductor as a function of temperature. (c) Specific heat of a soft superconductor.

currents die away to zero also. The complete Meissner effect shows that the superconducting to normal transition in a type I super-conductor is thermodynamically reversible and leads to the develop-ment of relations between the slope of the critical field–temperature curves and both the entropy difference and specific heat difference (for example, see Lynton, 1961).

The behavior of a soft superconducting body in a magnetic field can be conveniently described by attributing to it a magnetization cor-responding to a perfect diamagnetic. The ideal magnetization curve of a soft superconductor, supposedly a long thin cylinder parallel to the field, is shown in Figure 6.3. In practice, it has been possible to measure and produce such reversible curves. If the specimen is strained, or is in any way inhomogeneous, reversibility is lost; a typical return curve on decreasing H is shown by the dashed line, and now in zero external field the specimen shows a magnetic moment or trapped flux.

If the specimen is of a shape—e.g., a sphere—such that the Meissner effect itself upsets the magnitude of the external field, then the mag-netization curve will be modified, as shown by the dotted curve in Figure 6.3. At the external applied field where the local field at the surface of the superconductor first reaches the critical value, the speci-men passes into an intermediate state in which it subdivides into an alternating mesh of superconducting and normal regions·

6.1.1. Entering the surface of a soft superconductor, the magnetic field does not suddenly fall to zero but does so within a thin surface layer in which supercurrents can flow. The thickness of this layer is characterized by a distance λ known as the penetration depth. In the London formulation of the electromagnetic properties of a supercon-ductor, the magnetic field entering the plane surface falls off as $\exp(-x/\lambda)$, x being the distance in from the surface. In reality the field does not necessarily fall off in a simple exponential manner. The magnetization curves in Figure 6.3 must only apply to bodies of which the linear dimensions are very large compared with λ. Values

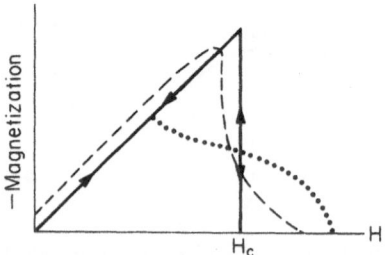

Fig. 6.3. Magnetization curve of a soft superconductor.

of λ usually lie in the range ~ 400 to ~ 1000 Å. If now the dimensions of the superconductor are much less than λ, the field penetrates throughout the body and the effective magnetization is very much reduced in magnitude. Consequently, when considering the free energy, G_s of the body in an external field H_e, given by

$$G_s(H_e) = G_s(0) - \int dv \int_0^{H_e} M(H_e) \, dH_e \qquad (6.1)$$

it can be seen that the value of H_e required to raise $G_s(H_e)$ to that of the normal material must be considerably increased. Thus the critical field to destroy superconductivity in a very thin film is enhanced roughly by a factor λ/t where t is the thickness. This leads at once to the idea that in superconductors which retain superconductivity to very high fields, the superconducting regions are very fine filaments much thinner than the penetration depth of the material concerned. Indeed Mendelssohn and Moore (1935) proposed a fine "sponge" of such filaments, which is discussed later. Superconductivity can also be destroyed by the magnetic field due to a current passing through the conductor. When the field at the surface reaches the critical value, the "critical current" is also reached.

6.1.2. The existence of the Meissner effect in a soft superconductor implies the existence of a positive surface energy. The argument is as follows: in an excluded external magnetic field H_e, the energy per unit volume of a superconductor increases by $\mu_0 H_e^2/2$. It would be energetically more favorable for the material to break up into alternating normal and superconducting layers such that the thickness of the superconducting layers is less than λ and the thickness of the normal material very much less still. The magnetic energy of the material could thus be heavily reduced. This type of behavior is unfavorable energetically if a positive surface energy E_s per unit area exists. If the material is assumed to be parallel-sided of thickness t, then it could not contain less than t/λ layers of thickness less than λ, and hence, allowing for the two sides of each layer, $(2t/\lambda)E_s > \mu_0 H_c^2 t/2$ or, on rearranging, $E_s > (\lambda/2)\mu_0 H_c^2/2$. It is customary to express the surface energy in terms of a parameter of dimensions of length such that

$$E_s = \Delta' \mu_0 H_c^2/2 \qquad (6.2)$$

and at once it follows that $\Delta' > \lambda/2$. One can see that should the surface energy become negative, then indeed in a magnetic field the material will not show a Meissner effect and will break up into a mesh of superconducting and normal domains.

6.1.3. In the London theory of the magnetic behavior of superconductors, the wave functions corresponding to superconducting electrons remain unchanged (are rigid) when an external magnetic field is applied. In his experiments on the variation of the penetration depth with temperature, Pippard (1950, 1952) pointed out that his results imply that the superconducting entropy is a function of magnetic field and that near T_c the changes of entropy with penetration depth can contribute as much as 25 % of the total entropy difference between the normal and superconducting phases. To assume that this entire entropy change takes place within the surface layer, which the London theory implies, results in an unreasonably high entropy density in this layer. Now in the Gorter–Casimir two-fluid model of a type I super-conductor, an "order parameter" ω is introduced, $0 < \omega < 1$. ω can be regarded as a measure of the fraction of free electrons in the super-conducting state. Pippard proposed that the order parameter ω can change slowly and significant changes can take place over a certain length ξ which he called the coherence length.

In the B.C.S. theory of superconductivity the coherence distance can be considered as a measure of the size of the "Cooper pairs." A change in the order parameter implies corresponding changes in the thermodynamic functions, and in this way the changes in entropy accompanying the imposition of an external field can be more widely distributed.

The surface energy is connected closely with the coherence length ξ. This may be shown with the aid of Figure 6.4, in which the variation of the order parameter and of the externally applied field H_e is shown in a direction perpendicular to the normal–superconducting boundary. Two "effective boundaries" to the solid can be defined. The dotted line M in Figure 6.4 represents a "magnetic boundary" given by supposing that inside the superconductor the flux B is $\mu_0 H_c$ up to M, where it drops sharply to zero going into the superconducting phase.

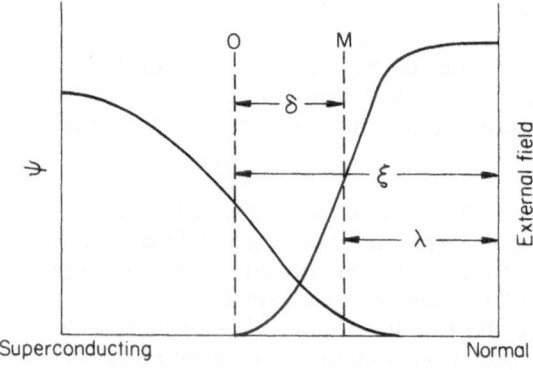

Fig. 6.4. Relation between λ and ξ for superconduction.

The total magnetic energy is $\int (BH/2)\,dv$ taken over the whole body, which can be equated to the actual value, so defining M. Likewise the configurational boundary O can be defined by supposing that the order parameter drops sharply to zero at O and equating the consequent free energy to the actual value. Suppose the distance OM is δ. The free energy per unit volume of a superconductor is less than that of the normal material by $\mu_0 H_c^2/2$. When boundary O lies inside M, as shown in Figure 6.4, it is equivalent to a reduction in the superconducting volume and so to an increase in the total free energy by $(\mu_0 H_c^2/2)\delta$ per unit area of the interphase boundary. The Pippard coherence range ξ implies a configurational boundary surface energy parameter $\Delta' v \xi$. From this surface energy the decrease in energy due to the field penetration must be subtracted $(\lambda \mu_0 H_c^2/2)$. Consequently, the net resulting energy parameter Δ is

$$\delta = \Delta \sim \xi - \lambda \tag{6.3}$$

6.1.4. In parallel with Pippard's ideas, Ginsburg and Landau (1950) proposed a phenomenological theory particularly designed to lead to a natural development of the surface energy and to a consideration of the destruction of superconductivity by a current. They introduced an order parameter ψ normalized so that $|\psi|^2 = n_s$, the density of superconducting electrons. In writing an expression for the free energy of the system in a magnetic field they added to the zero field energy the usual term $(H_e^2/2)$, where H_e is the external applied field, and also an extra term to take account of the possible appearance of a gradient in ψ. Thus ψ is not completely rigid with respect to the application of the external field, and we have the concept of a gradual extended variation in the superconducting order parameter analogous to that used by Pippard in proposing a range of coherence.

In discussing their results, Ginsburg and Landau introduced a dimensionless parameter κ defined as (cgs units)

$$\kappa = \frac{2e^{*2}}{\hbar^2 c^2} \cdot H_c^2 \cdot \lambda_c^4 \tag{6.4}$$

where

$$\lambda_c^2 = \frac{m_c^2}{2e^{*2}\psi^2}$$

and e^* was taken originally as the electronic charge, but later work shows that $e^* = 2e$ is better, corresponding to the Cooper pairs. In this theory κ, λ_c, the empirical penetration depth in the weak field limit, and H_c, the bulk critical field, can be thought of as parameters which are to be determined experimentally and which govern the properties of the material. It can be shown that in the limit $\kappa \to 0$, the external field ceases to affect the order parameter ψ and the penetration depth λ, and the Ginsburg–Landau model goes over to the original London model. An expression for the parameter Δ governing the surface energy can be developed but requires numerical integration. When $0 < \kappa \ll 1$, which seems to correspond to most of the simple pure elements such as tin and lead, $\Delta = 1.89\lambda/\kappa$, or Δ is of the order of $\sim 10\lambda_0$. (λ_0 is the empirical penetration depth in a pure superconductor). In later work, Bardeen (1954) and Gorkov (1959, 1960) have shown that the Ginsburg–Landau approach can be modified to be compatible with the Pippard coherence idea, and with the Bardeen–Cooper–Schriefer microscopic theories. Gorkov has shown that the parameter $\kappa \sim \lambda_0/\xi$, and thus the coherence length ξ and Δ are comparable in size. From Equation (6.2) it follows that when the range of coherence decreases and the penetration depth increases, Δ must decrease and eventually may become negative. In the Ginsburg–Landau approach, this occurs when $\kappa \geq 1/\sqrt{2}$. Abrikosov (1957) has treated this situation and gave the name "superconductor of the second kind" to those materials in which $\kappa \geq 1/\sqrt{2}$. This is frequently shortened to type II superconductors. The behavior of such materials in a magnetic field is represented in Figure 6.5, where the magnetization is shown as a function of H_c. Above a field H_{c_1}, the lower critical field, the material breaks into a mixed state compounded of a filamentary structure of normal and superconducting regions. Superconductivity is finally suppressed at the upper critical field H_{c_2}, which is well in excess of the field H_c corresponding to a type I superconductor. The larger κ is, the higher will be H_{c_2}. Goodman (1961) has also examined the situation by adding a surface energy term to the London expression for the free energy and has shown that, with a negative surface energy and a laminar normal–superconducting structure, behavior similar to Figure 6.5 is to be expected. As with the Mendelssohn sponge model, superconductivity

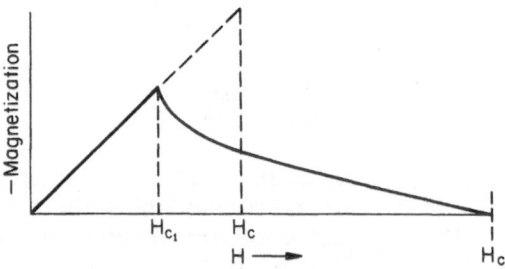

Fig. 6.5. Ideal magnetization of a type II super-conductor.

extending to high fields can be expected. In Figure 6.5 the magnetization curve should be reversible. There is now an extensive collection of experimental work which demonstrates this behavior. The alloys of indium and tin were examined by Davies (1960) and Wipf (1963). Livingston (1963) has examined a series of alloys and shows the way in which the type I curve for pure lead is modified as the proportion of the second component in the alloy is increased from zero (Figure 6.6). As the mean free path decreases, so ξ decreases and λ increases. Livingston was unable to show perfectly reversible curves and did not examine the current-carrying capacity of the material.

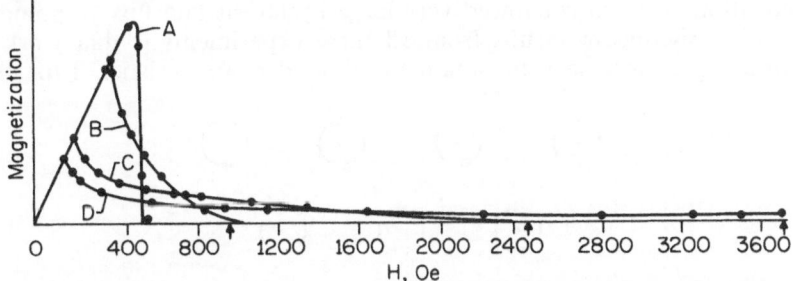

Fig. 6.6. Magnetization curves of annealed alloys of lead (after Livingston, 1963). (A) Pure lead; (B) lead + 2.08 wt. % In; (C) lead + 8.23 wt. % In; (D) lead + 20.4 wt. % In.

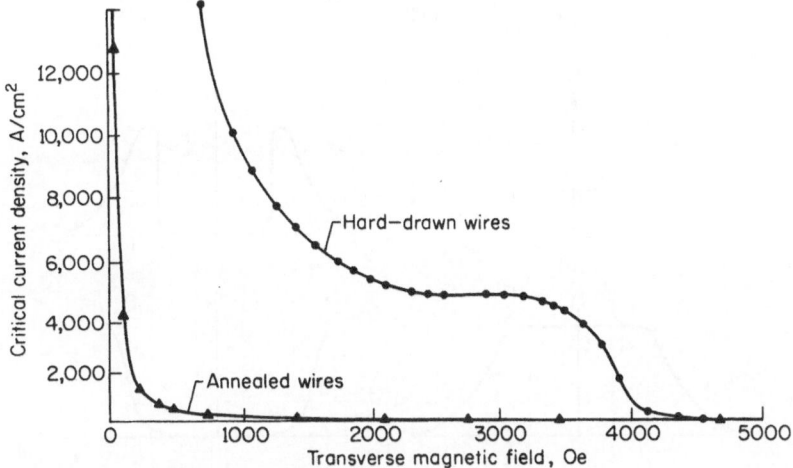

Fig. 6.7. Critical current *vs.* field curves for hard-drawn and annealed Ta -55 at. % Nb wires (after Heaton and Rose-Innes, 1963).

6.1.5. Heaton and Rose-Innes (1963) examined the alloys of niobium and tantalum and have produced significant results; such alloys are not very useful for building practical solenoids but are type II superconductors. Niobium and tantalum are members of the same subgroup of the periodic table and have atoms of very nearly the same size. They are miscible in all proportions. Using a carefully annealed 50–50 alloy, these authors were first able to show a magnetization curve which was virtually reversible. The corresponding current-carrying capacity is shown in Figure 6.7, and from our point of view is useless, for it falls to a very low value at quite low fields. On cold-drawing the alloy into a very hard-worked state, the current carrying capacity exhibited the desirable shoulder also shown in Figure 6.7. The magnetization curve now showed very large hysteresis and flux trapping.

The significant results from all these experiments is that work-hardening is necessary to obtain the desired characteristic. Type II

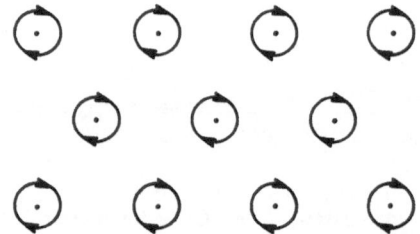

Fig. 6.8. Schematic representation of flux lines in a homogeneous type II super-conductor.

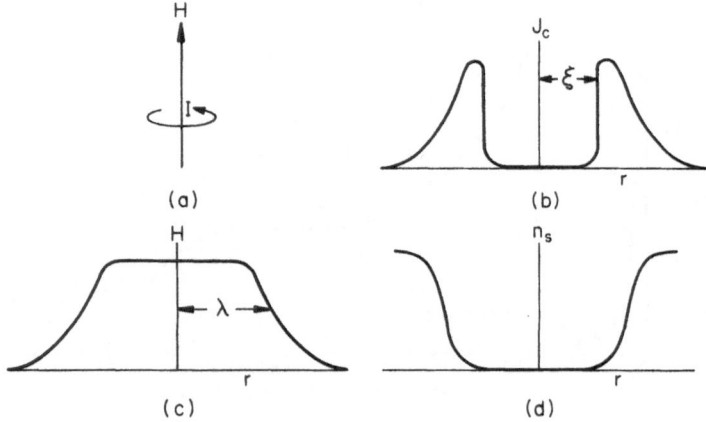

Fig. 6.9. Schematic representation of a quantized flux line.

superconductivity alone is not enough. The general behavior of the fully annealed specimen is what is to be expected from the Abrikosov model (1957). Quantized flux lines, units $\phi_0 = hc/2e = 2 \times 10^{-7}$ gauss/cm^2, can be imagined as uniformly penetrating the homogeneous material, each enclosed by supercurrent vortices and mutually repelling each other. This situation, occurring between H_{c_1} and H_{c_2} can be represented as in Figure 6.8 (flux normal to the paper). One can imagine the lines lying at the points of a uniform triangular lattice. Figure 6.9a represents a flux line with a circulating current I. In Figure 6.9b the supercurrent density is shown as a function of r, the distance from the flux line. The core of normal material can be imagined to be $\sim \xi$ wide but with somewhat ill-defined boundaries. This core contains most of the flux, but the field extends into the superconducting region over a distance $\sim \lambda$ as shown in Figure 6.9c. In Figure 6.9d the variation of n_s, the "superconducting electron density," is shown. It must be stressed that the mixed state which we are representing in this way is microscopic and in no way results from the distortion of the external field by the specimen; we are dealing with a specimen of zero demagnetization factor. If there are n lines per unit area, the average flux is $B = n\phi_0$, which can be taken as macroscopically uniform if the lines are uniformly distributed. B increases with H and the lines begin to "overlap" (Figure 6.9c), when $B \geq \phi_0/4\lambda^2$, at which point the distance between lines is approximately ξ (Figures 6.9b and 6.9d). The whole specimen becomes normal at the corresponding field $\sim \phi_0/4\xi^2$.

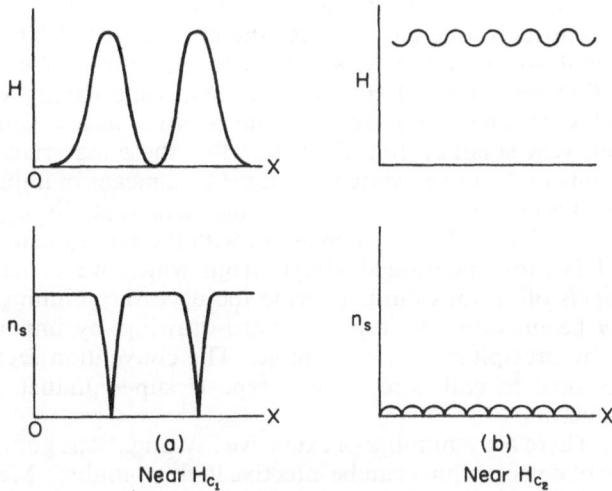

(a) Near H_{c_1} (b) Near H_{c_2}

Fig. 6.10. Variation of H and n_s inside a type II superconductor (a) near H_{c_1} and (b) near H_{c_2}.

Near H_{c_1} the microscopic variation of internal field and density of superconducting electrons can be represented as in Figure 6.10a, while near H_{c_2} they can be represented as in Figure 6.10b. In the microscopic supercurrent vortex around each flux line, the current density can reach 10^7 A/cm^2, but consideration of Figure 6.8 shows that the only net macroscopic current is at the specimen surface.

6.2. FLUX-PINNING AND CURRENTS IN TYPE II SUPERCONDUCTORS

6.2.0. For the present purposes, the point of interest is whether it is possible to have a dissipationless macroscopic "transport" current through the specimen. A transport current will disturb the equilibrium of the system by introducing two perturbations, the first because of the increase in free energy due to the kinetic energy of the transport current, and the second due to the interactions between the current and the field, the Lorentz force. When the current flows perpendicular to the magnetic field, as it does in a solenoid, the Lorentz force becomes the dominant perturbation. If the transport current is J taken along the y axis and the flux is along the z axis, there will be a gradient in B given by $\partial B / \partial x = -\mu J$ with a corresponding nonuniform flux line density B/ϕ_0. This cannot be a stable state, for flux lines will move along the x axis toward a configuration which will minimize the energy or to make

$$\frac{-\partial(B^2/2\mu)}{\partial x} = BJ = 0 \qquad (6.5)$$

Keeping I constant, one can see that the motion of flux lines should continue and so gives rise to an effective resistance. We therefore conclude that ideal type II superconductors, while having very high critical fields (H_{c_2}) have only zero resistance either at low fields below H_{c_1} or with very small current densities. For these materials to carry large currents in the mixed state there must be a means of impeding the motion of the flux lines or better pinning them spatially against the Lorentz force. These ideas are consistent with the experiments of Rose-Innes and Heaton mentioned above, from which we conclude that lattice defects of various kinds provide the necessary pinning centers. They may be introduced simply by hard-drawing, by imperfect sintering, or by precipitation, for example. The convention seems to be well-established to call hard-worked type II superconductors "hard superconductors."

6.2.1. There are a number of extensive investigations going on into the nature of defects which can be effective in flux pinning. Meanwhile, there has been a considerable advance in our understanding of some of the features of flux pinning as a result of several series of experiments notably by Kim and his co-workers (Kim *et al.*, 1963). Using hollow cylindrical specimens they were able to show that flux enters the specimen in jumps, particularly at low fields. From their studies of the magnetization of such specimens, Anderson (1962) has developed the idea of "flux creep."

Flux pinning is not "rigid," for, in spite of it, at finite temperatures flux lines will move through the material due to thermal activation. Figure 6.11a represents the uniform distribution of flux lines in an ideal type II superconductor. Physical inhomogeneities or defects responsible for flux-pinning can be represented as in 6.11b, where they are shown as small regions essentially nonsuperconducting, into which flux lines tend to crowd. The free energy in such regions is higher than that in the surroundings by $\mu_0 H_c^2/2$, and a free energy "modulation" across the specimen as in Figure 6.11c results. For motion of flux lines, such pinning centers become free energy minima and the average height of energy barriers for motion will then be $E_0 = (\mu_0 H_c^2/2)t^3 p$, where t is the average size of the defects and we suppose that only a fraction p of this barrier height is effective. Flux lines coming within a distance λ of each other must interact and will have to move more or less simultaneously to keep the local line density in equilibrium. As Figure 6.11b shows, the defects themselves aid the formation of flux "bundles." Thus in the presence of a transport current J, the Lorentz force can be taken as acting on a whole bundle of flux lines, and is given by $JB\lambda^2 d$, where d is the mean distance between defects in the direction of flux lines. Such a force tilts the barrier structure of Figure 6.11d to a configuration as in Figure 6.11e and the mean barrier energy is reduced to

$$E_p = \frac{p\mu_0 H_c^2 t^3}{2} - JB\lambda^2 d$$

$$= E_0 - q\alpha$$

(6.6)

where $\alpha = JB$.

Thus at a finite temperature T, flux lines can be expected to "migrate" over the barriers in bundles and in a "hopping" process with a rate which will be proportional to $\exp[-(E_0 - q\alpha)/kT]$. The velocity for flux migration or creep will then be $V = V_0 \exp[-(E_0 - q\alpha)/kT]$. Although V_0 cannot be estimated with any precision, the exponential factor dominates the situation. Flux lines will always move in the presence of a transport current (i.e., for a finite value of α), but they may do so undetectably slowly. Such an exponential decay has been observed in a number of experiments on cylinders and wires by Kim *et al.* (1963), and these ideas are fairly well established. Nevertheless, the creep rates are sufficiently slow to regard flux-pinning in a hard superconductor as stable for practical purposes.

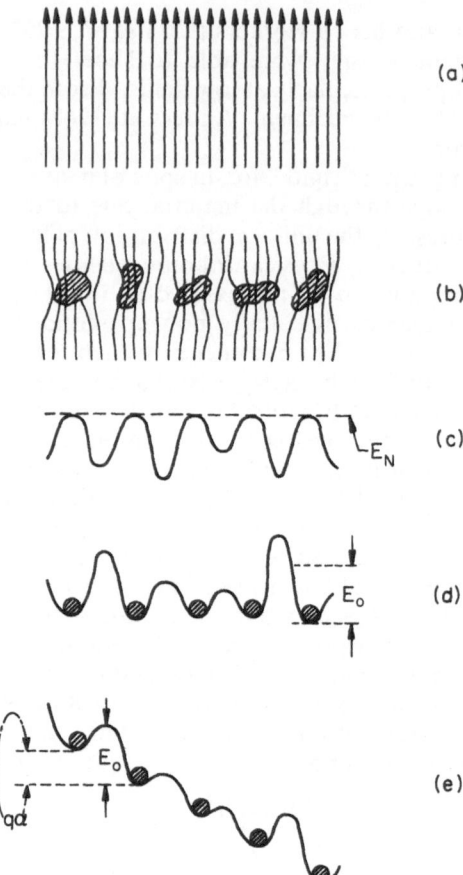

Fig. 6.11. Effect of physical inhomogeneities in flux lines, between H_{c_1} and H_{c_2} (after Kim, 1964). (a) Uniform field and homogeneous material giving uniformly spaced flux lines; (b) formation of "bundles" due to "inhomogeneities"; (c) free-energy modulation; (d) barrier height; (e) hop rate $\sim \exp[(-E_0 - q\alpha)/KT]$.

6.3. FURTHER PROPERTIES AND LIMITATIONS OF TYPE II SUPERCONDUCTORS

6.3.0. In Sections 6.1 and 6.2 the basic ideas concerning type II superconductors have been set out. There are a number of other results which are relevant to the use of these materials in "supermagnets."

First, there are a number of experimental results which show that in type II materials which contain defects, changes in flux density due to changes in the external field often occur in "jumps," in which whole "domains" change their state of magnetization. The most graphic results are sets of photographs (and films) due to Desorbo and Healy (1964), in which the state of magnetization of tantalum sheets is shown by a reflection technique using plane-polarized light passing through a sheet of cerium glass placed on the specimen.

The basic ideas of pinning due to a "modulation" in free energy carried by the defects introduced by hard work can be used to explain the highly anisotropic critical current densities (ratios as high as 50 : 1) which are observed in cold-rolled strips of niobium–titanium alloys, for example. In such materials the critical current density is a maximum when the Lorentz force is directed perpendicular to the flow lines in the material which are revealed by etching. The Lorentz force is then in the direction of maximum free-energy modulation.

A recent development has been the theoretical prediction by Saint-James and de Gennes (1963) of a superconducting surface sheath which should exist to fields higher than H_{c_2}. Superconductivity should be observed due to this sheath up to a further limit H_{c_3}, but between H_{c_2} and H_{c_3} the core of the material should be normal. This effect has now been demonstrated. In Figure 6.7, showing the curves of critical current against field for the niobium–tantalum alloys investigated by Rose-Innes, a tail to the curve very near to the horizontal axis can be seen. This could be due to the existence of a superconducting sheath as proposed by de Gennes. These results can be summarized in Figure 6.12, a "phase diagram" in which κ is plotted as a function of $h = H/H_c$, the reduced magnetic field. Thus the behavior expected of a material of a given κ as the field increases can be found by tracing along a horizontal line corresponding to that κ.

6.3.1. In the general development of the Abrikosov theory the upper critical field H_{c_2} can be expressed in terms of the various electronic parameters of the material. Thus,

$$H_{c_2} \sim [\rho_n \gamma T_c + A\gamma^2 T_c^2 (S_F/S)^2 n^{4/3}] f(t) \tag{6.7}$$

where $f(t)$ is a function containing even powers, up to the sixth, of the reduced temperature T/T_c. ρ_n is the normal state electrical resistivity, γ the electronic specific heat, n the average number of electrons outside closed shells per unit volume, S is the actual area of the Fermi surface and S_F is the Fermi surface corresponding to a free-electron gas of density n, and A is a numerical constant. Such an expression can be useful as a guide in the search for materials of higher H_{c_2} than those which at present exist. One can see at once the interest in the transition metals with their high values of ρ_n and γ.

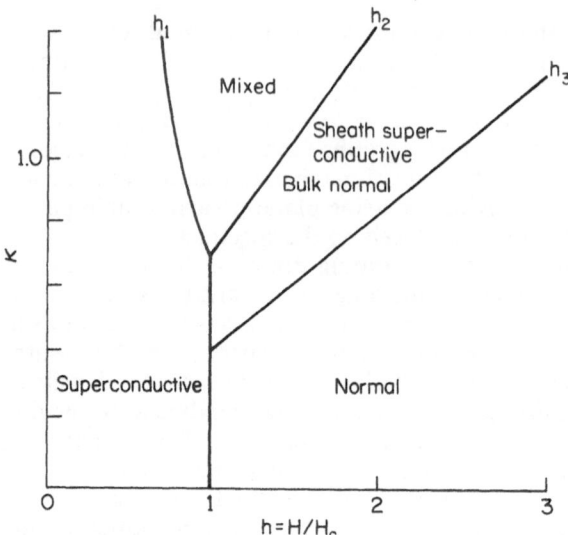

Fig. 6.12. "Phase diagram" for type II superconductors.

6.3.2. However, Clogston (1962) and Chandrasekhar (1962) have independently pointed out that there is a limitation on the upper critical field which arises from the action of the externally applied field on the electron spins in the normal material. When the latter begin to line up in very powerful fields, the free energy of the normal material begins to be reduced, while that of the superconducting materials is rising. The expression of Equation (6.7) was deduced on the assumption that the normal material free energy was field-independent, so it must, therefore, be unduly optimistic. The whole material will become normal at fields where the free energy of the normal material is just lower than that of the superconducting. If the Pauli paramagnetic energy is equated to the zero-field energy difference between the normal and superconducting states, the following relations can be derived for the "paramagnetic" limited upper field:

$$H_p = 18.4 \times 10^3 \, T_c(1 - t^2) \quad [\text{Oe}] \tag{6.8}$$

Berlincourt and Hake (1963) have produced a series of results supporting these ideas. They examined a series of alloys between vanadium and titanium in which the electron/atom ratio varied from near 4 to near 5. Their results are shown in Figure 6.13. H_r is the measured field at which the resistance could just be observed with a current density of 10 A/cm² at 1.2°K. It corresponds very closely to the intercept of the

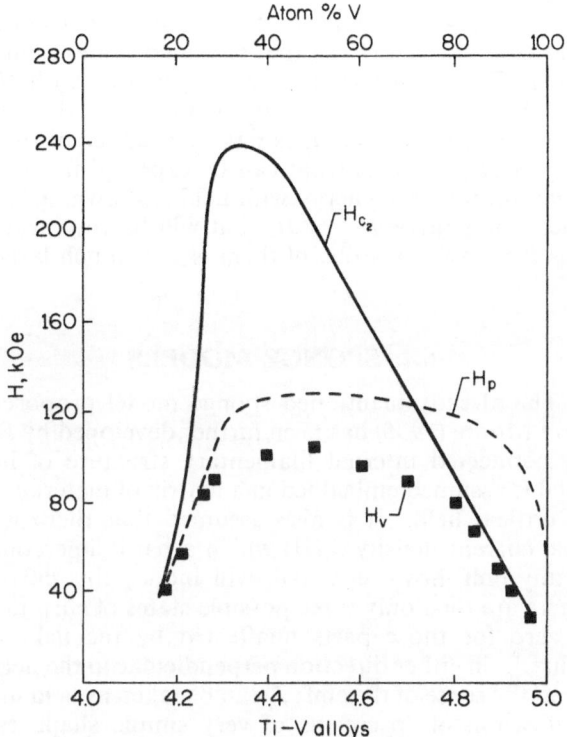

Fig. 6.13. Critical fields for Ti–V alloys (after Berlincourt and Hake, 1963).

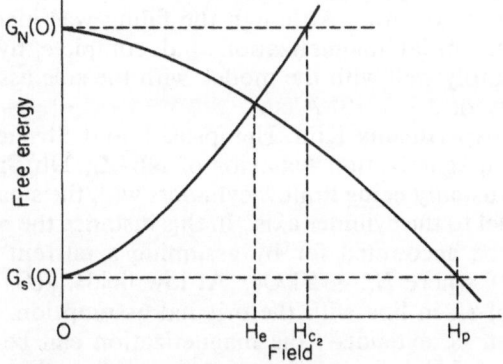

Fig. 6.14. Free energy as a function of field showing the effect of paramagnetic limit. H_e—experimental upper critical field; H_{c_2}— calculated upper critical field, H_p—paramagnetic limit.

extrapolated sharply falling part of the critical current/field curve with the field axis. H_{c_2} and H_p are the calculated values using Equations (6.7) and (6.8). The experimental results agree well with H_p and H_{c_2}, whichever is the least, at either end of the range. In the middle of the range of electron–atom ratio, H_r is less than either of the calculated values of H_p or H_{c_2}. This is what can be expected from Figure 6.14, where the variation of free energy with field is shown. Clearly H_r will always be less than either H_{c_2} or H_r, but will be very nearly equal to whichever is the least when one of them is very much larger than the other.

6.4. SPONGE MODELS

6.4.0. The already mentioned sponge model proposed by Mendelssohn and Moore (1935) has been further developed by Bean (1964). A multiply connected internal filamentary structure of high critical field material is assumed embedded in a matrix of material of lower (or even zero) critical field. It is also assumed that there is a limiting macroscopic current density $J_c(H)$ which a hard superconductor can carry and any emf, however small, will induce this current to flow locally. There are then only three possible states of current flow in the specimen; zero for those parts unaffected by the field, or the full limiting value J_c, in either direction perpendicular to the field direction, depending on the sense of the emf; J_c can be taken as field independent. The magnetization of specimens of very simple shape can then be calculated using this idealized model. In spite of the crudity of the assumptions, Bean, Doyle, and Pincus (1962) have applied it to a system made from porous Vycor glass into which lead had been forced at 335°C and 0.3×10^4 atm. Although the filling was incomplete, they found that the initial magnetization and complete hysteresis loop agreed remarkably well with the model, with the sole assumption of a current density of 2.7×10^4 A/cm^2.

In other experiments Kim, Hempstead, and Strnad (1963) have examined the magnetization behavior of Nb–Zr, Nb$_3$Sn, and V$_3$Ga, the specimens usually being hollow cylinders with the steady magnetizing field parallel to the cylinder axis. In this instance the magnetization curves could be accounted for by assuming a current density $J(H)$ $= \alpha/(H^* + H)$, where $H^* \sim 5$ kOe. At low fields, $J(H)$ is practically independent of H, in line with the original assumption. This kind of experiment can be extended: the magnetization can be examined in strong steady fields by superimposing a small collinear sinusoidal component. Thus the specimen traverses minor magnetization loops and the response, which will be nonsinusoidal, picked up in a small secondary coil. Examination of the harmonic content in the secondary gives sufficient information to deduce the limiting macroscopic current density J_c, although the ratios of the experimental amplitudes of the various harmonics do not agree with theory. Noise apparent in such experiments can be attributed to small flux jumps, presumably due to flux breaking through individual threads of the matrix (compare the origin of Barkhausen noise).

Parallel series of experiments have been carried out by Seraphim, d'Heurle, and Heller (1962) on the Pb–Al composites, likewise the Th–Nb eutectic has been examined by Cline, Rose, and Wulff (1963).

6.5. TYPE II SUPERCONDUCTORS AND SPONGE MODELS

6.5.0. In our discussions in Section 6.2, on type II superconductors, we considered the motion of flux lines in homogeneous material and their pinning in inhomogeneous material. The driving force to push flux lines into inhomogeneous material derives from the increase in external field. Once a flux line has penetrated into the interior of the material, the driving force is supplied by interaction with other flux lines.

If the local average internal field is $B = \phi_0 n$, where n is the density of flux lines, the force per unit length on a flux line can be shown to be (Friedel *et al.*, 1963):

$$F_m = -\left(\frac{\phi_0}{\mu_i}\right)\frac{\partial B}{\partial x}$$

where

$$\mu_i = \frac{dB}{dH}$$

on the ideal reversible curve at the value B. At the absolute zero a flux line will remain pinned until the magnetic driving force becomes just greater than the pinning force F_p created by defects. The condition $F_p = F_m$ defines a critical flux gradient given by

$$\left(\frac{\partial B}{\partial x}\right)_{\text{crit}} = \left(\frac{\mu_i}{\phi_0}\right)F_p$$

In an increasing field, flux will continue to move into the specimen until the flux gradient is everywhere just below this critical value. The distribution of internal flux and thus the specimen magnetization can be specified in terms of $(\partial B/\partial x)_{\text{crit}}$ as a function of B and the boundary conditions for B at the surface. From Maxwell's equations macroscopic critical flux gradients are equivalent to macroscopic "critical currents." This gives a means of relating the flux-pinning models outlined in Section 6.2 to the sponge model of Section 6.4. In the Bean sponge model the critical current density J_c was taken as independent of field, with the boundary condition at the surface that $B = \mu_0 H$. If the restriction on the current density is removed and the surface B is taken as equal to some value in equilibrium with the external field, a rather more general model results which corresponds to the flux-pinning model already discussed.

6.6. CHARACTERISTICS OF HARD SUPERCONDUCTING MATERIALS FOR SOLENOID CONSTRUCTION

6.6.0. In this section materials are discussed which have been used, or which could be useful in the future, for solenoid construction. They are dealt with more or less in ascending order of critical field. In all instances where critical current–field curves are given, they refer to *short–sample* tests. These materials fall into two classes, alloys and compounds. All the alloys discussed are sufficiently ductile to make probable the production of very long lengths of hard-worked material, whereas all the compounds of interest are of the β tungsten structure and are both hard and brittle. They cannot be drawn and worked so other means of preparation have to be resorted to, such as those of Section 6.6.5.

6.6.1. Niobium. Yntema (1955) first used niobium wire on a small iron-cored electromagnet. He showed that his wires could carry current densities of 7.4×10^4 A/cm^2 at 5 kOe. Similar results have been produced by Autler (1960) and by Kunzler (1961).

6.6.2. Molybdenum–Rhenium Alloys. Alloys of this system were investigated by Kunzler (1961), using short samples of thickness ranging from a few hundredths of a millimeter to about 1 mm. They produced the results shown in Figure 6.15 for the alloy of composition Mo$_3$Re, (diameter 0.007 cm). The desirable knee-shaped characteristic is well shown. At a point such as A in Figure 6.15 the critical current density is 0.65×10^4 A/cm^2. An experimental coil, 2 cm in diameter by about 3 cm long, wound from wire of this material, 0.007 cm in diameter (30,000 turns), gave the results shown in Figure 6.16. The lower line corresponds to all the wire used in series. By splitting the solenoid so that the central part was used in parallel, rather higher fields (corresponding to the upper line) could be produced. The critical current behavior of such coils does not seem to differ appreciably from that which one would calculate from the short sample test results shown in Figure 6.15.

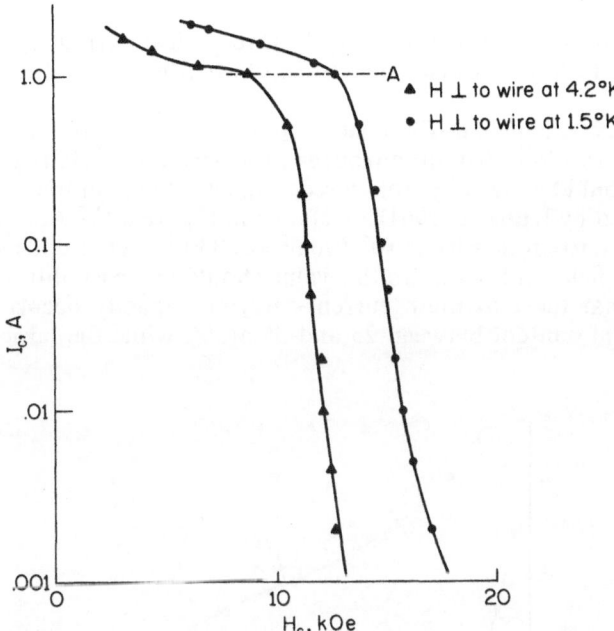

Fig. 6.15. Critical field curves for Mo₃Re (after Kunzler, 1961).

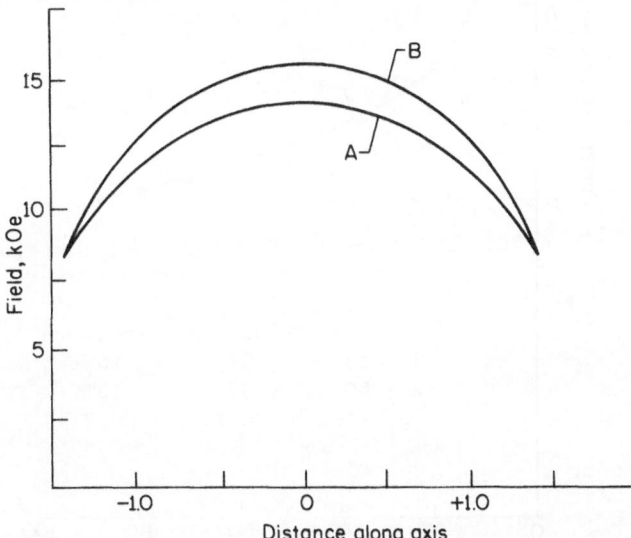

Fig. 6.16. Field on axis of Mo₃Re coil with windings in series and in parallel (after Kunzler, 1961). (A) Windings in series; (B) Windings in parallel.

Alloys of composition Mo_2Re have critical current curves going to about 10% higher fields than those of Mo_3Re.

6.6.3. Niobium–Zirconium Alloys. Alloys of this system were the first to offer striking advances in generating fields beyond those which could be given by iron-cored magnets. Some of the early results published by Kunzler (1961) are shown in Figure 6.17. At a point such as B the current density is 10^4 A/cm² at 80 kOe. Thus coils which can produce fields at least up to this value should be possibilities. Kunzler found that the maximum current-carrying capacity occurred with a zirconium content between 25 and 35 at. %, while the largest critical

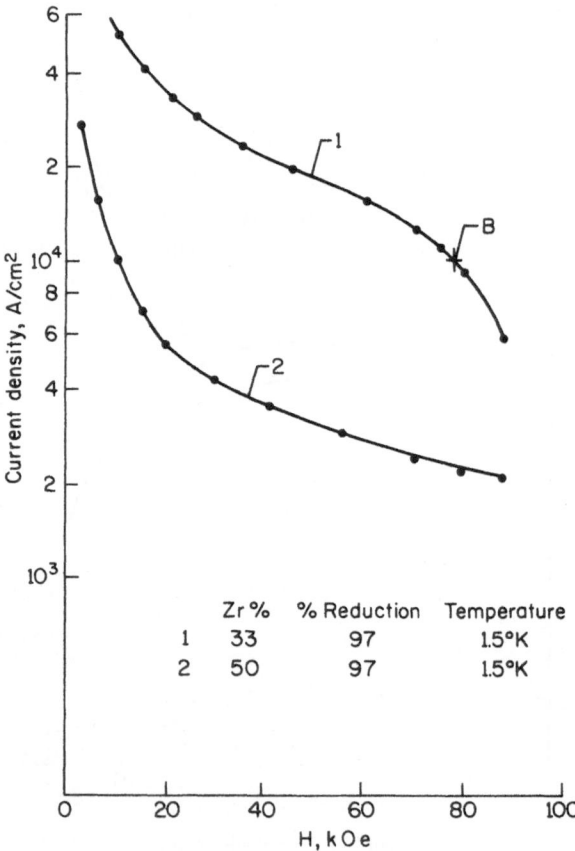

Fig. 6.17. Critical current *vs.* field curves for Nb–Zr alloys (after Kunzler, 1961).

field occurred at compositions between 65 and 75 at. % zirconium. All specimens were heavily cold-worked (97% reduction).

These alloys are ductile and, provided the purity of the material is good, very long lengths can be drawn and a high degree of cold work used. There are now several commercial sources both in the USA and in Europe, though wires of nominally the same composition differ somewhat according to the amount of cold work and the annealing recipe used during manufacture. Thus annealing at 700 to 800°C for 30 min appears to improve the maximum critical current as given by short sample tests, for wire of composition Nb–25% Zr.

Wires made from these alloys form the basis of fully stabilized superconducting strip described by Stekly and Zar (1965) and discussed fully in Sections 7.4 and 7.5.

6.6.4. Niobium–Titanium Alloys. These alloys are of great interest. Like the niobium–zirconium system they are ductile; moreover, their upper critical fields are considerably higher (see Berlincourt and Hake 1962, 1963), but the critical current densities so far reported are smaller. In Figure 6.18 a critical current/field curve is shown for

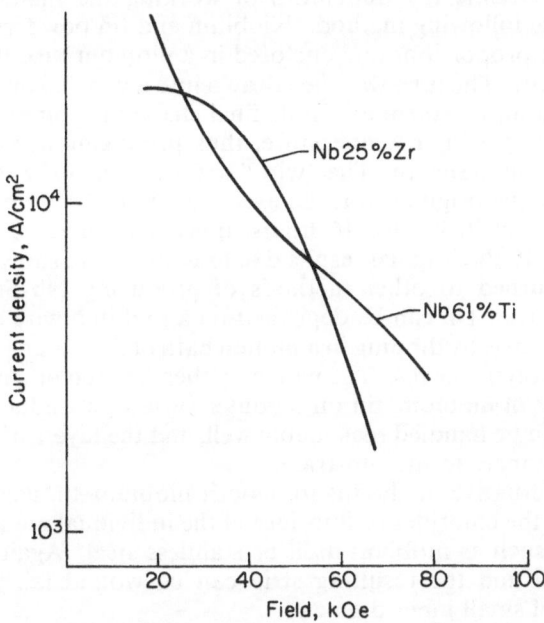

Fig. 6.18. Comparison of NbZr and NbTi alloys (after Coffey, Hulm, et al., 1965).

comparison with that of Nb–25 % Zr. Obviously such wire is very useful for the inner windings of a magnet designed to generate fields near 100 kOe and indeed it has been so used by Coffey, Hulm, *et al.* (1965).

Recently Vetrano and Boom (1965) have reported that by suitable metallurgical treatment an increase in the critical current density can be achieved to greater than 10^5 A/cm^2 at 4.2°K and in 30 kOe for an alloy of composition Ti–20 at. % Nb. It should be possible to use these alloys in the same way as those of the Nb–Zr system in fully stabilized solenoids.

6.6.5. Niobium–Tin or Niobium–Stannide, Nb_3Sn.

Niobium–tin, Nb_3Sn, is a compound of β tungsten structure. The first results showing its promise were published by Kunzler (1961). Besides carrying high critical currents at high fields (Figures 6.19 and 6.20), the upper critical field measured in steady fields, as a function of temperature, rises steeply, as shown in Figure 6.21. Linear extrapolation suggested a value of about 350 kOe at 0°K. Recently specimens have been examined at the National Magnet Laboratory, M.I.T., and the upper critical field found to be 220 kOe at helium temperature (Montgomery 1966).

To overcome the difficulties of working the material, Kunzler adopted the following method. Niobium and tin powders were mixed in the right proportions and enclosed in a niobium tube (bore 3.2 mm, o.d. 6.35 mm). The tube was then drawn into a wire 3.5 mm in diameter and wound into a solenoid. In its final shape the complete coil could be sintered at a high temperature, thus producing a core of Nb_3Sn within a niobium sheath. The "wire" so formed is too brittle to wind or unwind. In the original work an excess of about 10 % tin with a firing temperature of 970°C for 16 hours appears to have been successful. In Figure 6.19 the original results due to Kunzler are shown. Attention has now turned to other methods of producing Nb_3Sn in a more tractable form. Tin can be deposited on a niobium wire directly from the vapor phase, by dipping in a molten bath of tin under a suitable flux, or by electrolytic means. The wire may then be fired or sintered to give a thin layer of niobium–tin on its outer surface. Conductors made in this way can be handled reasonably well, and the layers of niobium–tin appear to adhere to the substrate.

An alternative method is to deposit niobium–tin itself by decomposition of the chlorides or fluorides of the individual metals on heated substrates such as niobium itself or stainless steel. Again such layers adhere well and the resulting strip can be wound fairly easily into solenoids of small inner diameter.

A wide range of widths (2–5 mm) and very long lengths seem possible by this method of production. Such material can be obtained

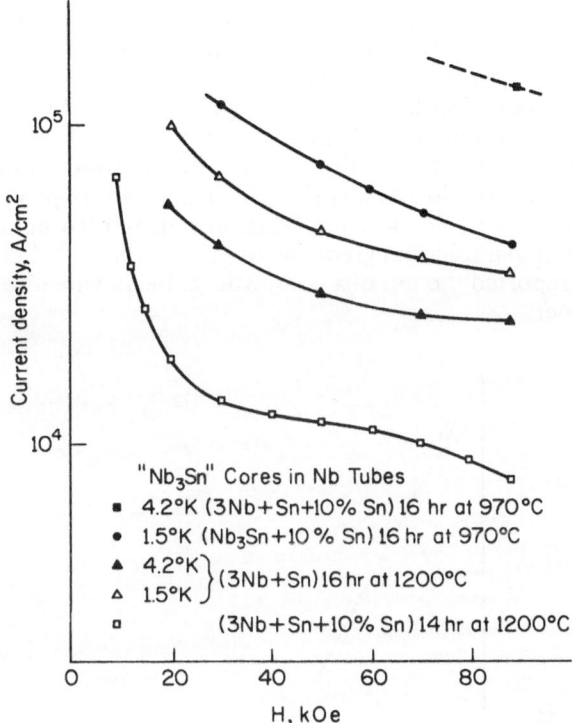

Fig. 6.19. Critical current *vs.* field curves for Nb₃Sn
(after Kunzler, 1961).

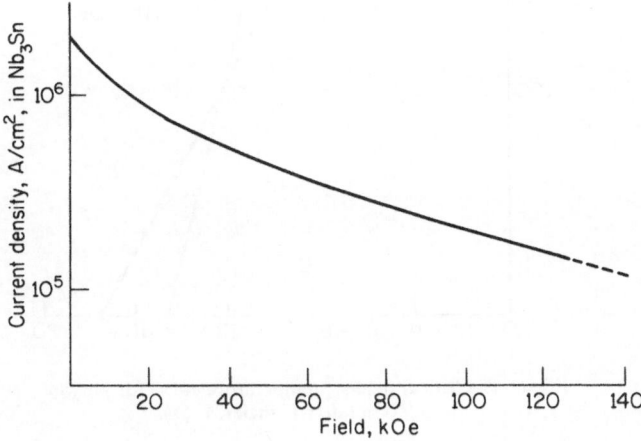

Fig. 6.20. Typical critical current *vs.* field curves for Nb₃Sn strip.

commercially* usually in widths of 2.3 mm and thickness 0.08 mm and coated with silver. Material produced in this way has been used to produce fields of about 112 kOe (Sampson, 1965). The tin-coating method of production has been developed particularly in the form of multistranded conductors. Benz, Martin, and Bruch (1965) describe the details of preparation and give much information on the performance of this material in small coils. The usual form of production is in 7- or 49-stranded cable. Martin, Benz, Bruch, and Rosner (1963) used such material in a solènoid giving fields greater than 100 kOe. Cornish (1966) has reported the use of a composite cable in which some strands are of copper.

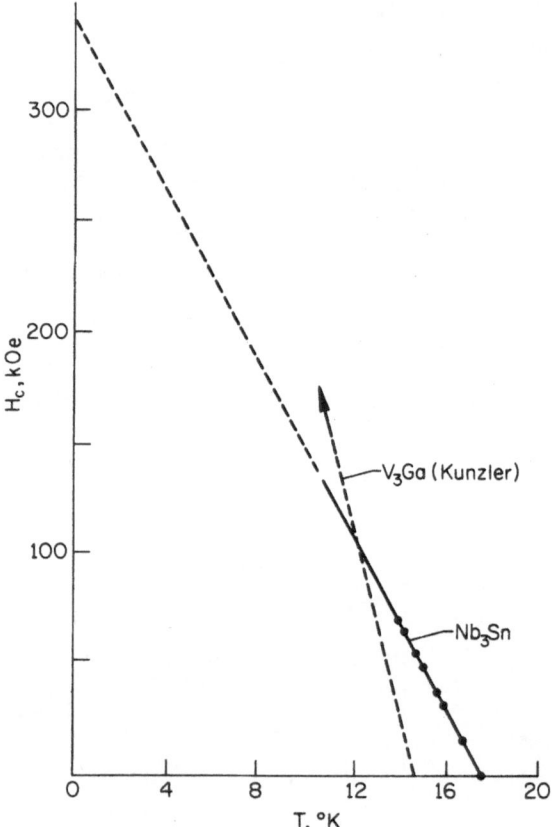

Fig. 6.21. Critical field *vs.* temperature curve for Nb₃Sn and V₃Ga (after Kunzler, 1962).

* For example, from RCA Electronic Components and Devices, Harrison, N.J.

6.7. OTHER HARD SUPERCONDUCTORS AND FUTURE MATERIALS

Equation (6.6) gives the expression for the paramagnetic limiting field as proposed by Clogston. This shows that materials which naturally have a high value of H_{c_2}, as derived from the Abrikosov theory, have high values of T_c, which focuses attention on the transition metals and their compounds or alloys with their high values of G_n and γ. It is, therefore, not surprising to find that the most successful materials so far found are the niobium–zirconium alloys and niobium–tin. Two other materials of high T_c which seem to be interesting are vanadium silicide, V_3Si, and gallium vanadide, V_3Ga; both have transition temperatures near $18.5°K$ and both are of the β tungsten structure.

Kunzler (1961) has reported that composition $V_{2.95}Ga$ has a critical field temperature curve as shown in Figure 6.21, where it is compared with that of Nb_3Sn. Again a linear extrapolation suggests the possibility of very high critical fields. This material and vanadium silicide must be the objects of much further investigation to see if they can offer any advantage over Nb_3Sn as far as current-carrying capacity is concerned. Both can probably be prepared by vapor-decomposition techniques. Whether there is much significance to be attached to the β tungsten structure as such in this connection is difficult to say. The structure itself is interesting: it is cubic, class A.15, and Figure 6.22 shows a cell for a compound A_3B. Those atoms designated by A are arranged in orthogonal nonintersecting linear chains. Their nearest neighbors are the two adjacent A atoms. In Nb_3Sn the Nb atoms are some 10–15% nearer in their chains than they are in the pure metal. The B atoms have 12 nearest A atom neighbors.

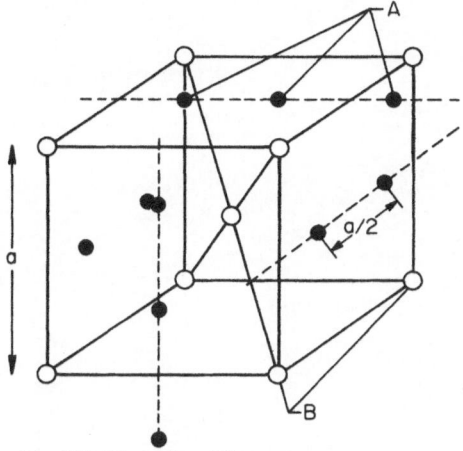

Fig. 6.22. The ordered β-tungsten structure.

Chapter 7

Superconducting Solenoids

7.0 INTRODUCTION

In this chapter the situation concerning the design and construction of superconducting solenoids is reviewed. It is natural to begin with a simple theoretical analysis. However, there are a number of practical factors which must be taken into account if an analysis is to be more than an exercise referred to ideal circumstances. These can be grouped under three headings:

(i) *Degradation factors:* The results of short sample tests for the critical current–magnetic field relationship for hard superconductors can be misleadingly optimistic if used to forecast the behavior of the wire when wound into a solenoid, unless the wire is suitably clad with a sufficient thickness of a good conductor. In the worst cases degradation of performance by factors as large as two or three or more can occur.

(ii) *Protection of the coil from inadvertent quenching* (returning to normal resistivity): The energy stored in a magnetic field of 100 kOe is about 40 J/cm^3; if the solenoid is of significant size (e.g., 100 cm^3) and the wire becomes normal at a particular point, the magnetic energy will be dissipated there as Joule heating. The quantities involved are quite sufficient to fuse the wire and destroy the coil locally.

(iii) *Energizing the coil:* Even quite thin superconducting wires can carry supercurrents of many tens of amperes. Solenoids made from thicker conductors are easier to build and on inadvertent quenching the voltages built up in them will be smaller, but the introduction of large currents into a cryostat raises a number of problems.

The bulk of this chapter will be concerned with these three points and with a discussion of present solutions to the problems raised.

7.1. THEORETICAL ANALYSIS

7.1.0. In a solenoid employing superconducting windings the electrical power needed is already zero, and thus the power optimization analysis of conventional coils is no longer relevant. However, superconducting materials are expensive; also energy roughly proportional to the coil volume must be expended in cooling down the solenoid. The optimization required is that of obtaining the desired field with the smallest possible quantity of superconducting material. The treatment given here is based on work by Boom and Livingston (1962), Gauster (1962), Wood (1962b), and Thomas and Bright (1966).

Hard superconductors are available in wires or strips of constant cross section, the most usual forms having already been summarized.

For the moment let us ignore degradation effects. In a square-ended solenoid, wound with uniform space factor, we have already seen that the field at the center is given by

$$H_0 = K r_0 J_0 \lambda. \tag{7.1}$$

If then A is the cross-sectional area allowed for each conductor in fabricating the coil, the current per conductor is given by the product of this area and the effective current density, $J_0 \lambda A$. It is convenient to introduce a volume factor, v, given by

$$v = 2\pi\beta(\alpha^2 - 1) \tag{7.2}$$

whence the conductor volume is $r_0^3 v$ and the conductor length equals $r_0^3 v/A$. On substituting in Equation (7.1) we find

$$H_0 = K r_0 I/A \tag{7.3}$$

In the usual case of uniform current density the Fabry factor, G, is introduced, for charts of it plotted as a function of α and β have already been drawn up and published. Thus Equation (7.3) becomes

$$H_0 = G\sqrt{v}r_0 I/A = G\sqrt{lI^2/Ar_0} \tag{7.4}$$

It is usual to display contours of constant G and lines of constant v. The (G_{max}, v_{min}) curve is also drawn, points on it giving the maximum G and the corresponding values of α and β for a minimum v or conductor length. Graphs of $G\sqrt{v}(=$ our $K)$ taken on the $G_{max} v_{min}$ curve as functions of α and β have also been published. (Note that the published figures are in cgs units.)

In a first approximation a current density may be chosen, from the known wire characteristics, as the maximum which the wire can carry in a desired field, while the space factor λ is determined from the form of winding and insulation, as is A. Thus for a given r_0 one can

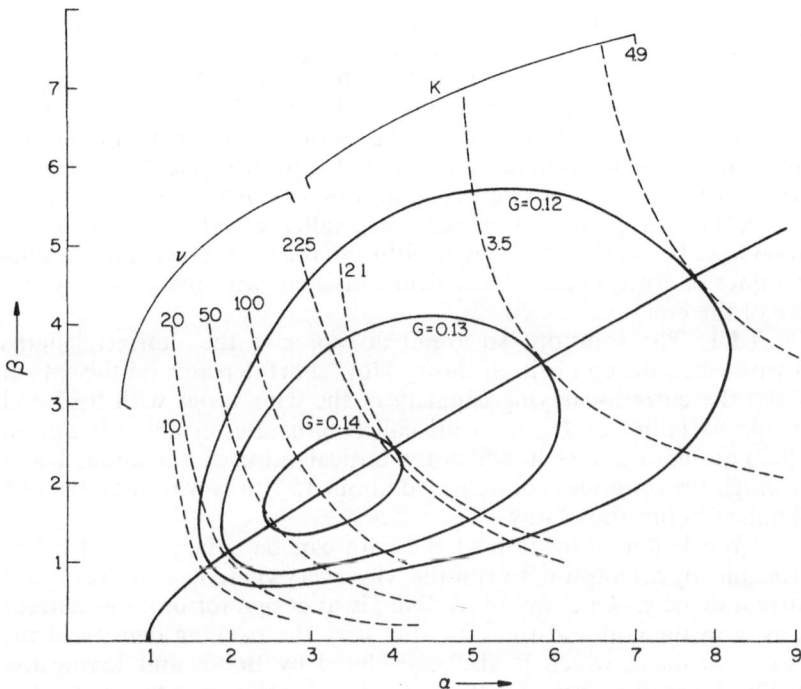

Fig. 7.1. Contours of constant G, K, and v as functions of α and β.

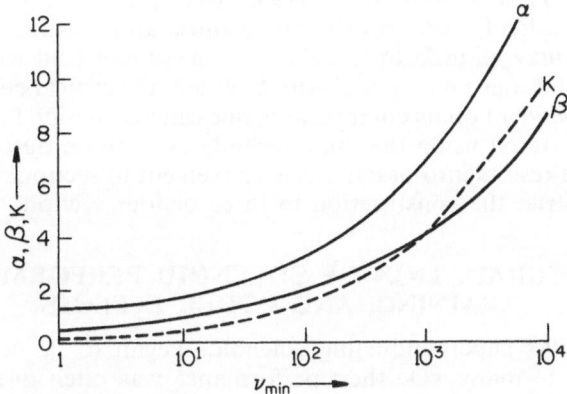

Fig. 7.2. α, β, and K as functions of v_{min}.

arrive at a value of $G\sqrt{v}$ and from Figures 7.1 and 7.2, the corresponding values of G_{max}, v_{min}, α and β, which thus gives an idea of the form of coil for minimal length. But the field at the center of the coil is less than the maximum, H_m, in the windings, which is usually to be found at the inner turns on an equatorial plane. The ratio H_m/H_0 is already known and plotted as a function of α and β, and it is H_m which is the quenching field. From this first trial a new (and lower) value of current density can be chosen so that it matches H_m (alternatively a value of H_m could have been guessed to begin with) and the process repeated aiming at values of critical current and field consistent with the geometry and size of the coil.

7.1.1. The solutions so found do not give the shortest lengths of wire when the coil is itself short. How short depends on the rate at which the current-carrying capacity of the wire varies with field and on the variation of H_m/H_0 with coil length. Boom and Livingston (1962) quote examples in which the critical value of β is about 1 and in which the optimum coil required about 15% less wire than the one obtained in the above way.

Even better utilization of the wire can be achieved, at least in principle, by attempting to run the wire everywhere near to its critical current density. One way of so doing is to design for uniform current density in the coil and then suitably vary the packing density of the wire, a problem which is also considered by Boom and Livingston (1962). A practical way of getting near to this solution is to sectionalize the coil and to run each section at a different current. The optimization problem between each section is then more difficult and has been considered by Gauster and Parker (1962). The potential saving of wire is great and could be up to 40%.

7.1.2. Thomas and Bright (1966) have published a variation of these methods. Procedures similar to those above utilizing the same equations may be used to calculate the maximum field which can be generated for a given length of wire l. Where the critical current of the wire is dependent on its composition, one can also design for minimum mass of material using the same techniques as those outlined above. To put the results into practice it is convenient to sectionalize the coil and to restrict the construction to three or four sections.

7.2. DEGRADATION OF SOLENOID PERFORMANCE, TRAINING, AND OTHER EFFECTS

7.2.0. As superconducting solenoids began to be developed to give fields of many kOe their performance was often disappointing. The critical current of the wire when wound into a solenoid was often less than half that given by short sample tests and the anticipated high fields were not generated. This behavior was noticeable with niobium–zirconium alloys and particularly with the niobium-rich materials, of which a typical example is shown in Figure 7.3. The curve *A* gives the short sample test results, while *BB* gives the critical current–field curve for a series of solenoids wound from the same wire (Hulm, Chandrasekhar, and Riemersma, 1963). For a given solenoid the field is linearly related to the current as is represented by the line *OX* in Figure 7.3 which cuts *AA* at *X*, the hoped-for operating point for the innermost turns of the coil. In practice the operating point was *Y* on *BB*. A "training effect" could also often be observed, in which the

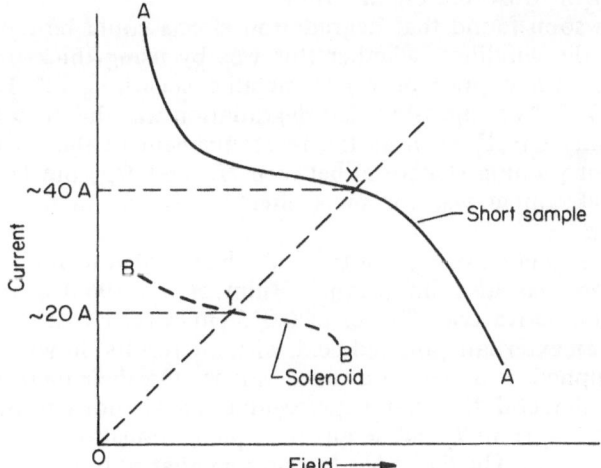

Fig. 7.3. Degradation in a superconducting solenoid.

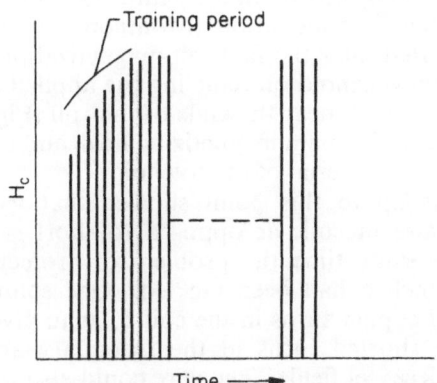

Fig. 7.4. Training effect in an Nb–Zr
solenoid.

coil had to be energized for a number of cycles to obtain the best performance. A typical example is shown in Figure 7.4.

7.2.1. Of the explanations which began to be put forward, the first suggestion was that there were "weak spots" in the wire. When lengths of thousands of meters are drawn, the state of work-hardening must vary along the length of wire. There could also be a lack of homogeneity or a segregation of impurities in different parts of the original ingot from which the wire was drawn. Differences in the critical current–field curves for each section of the wire can, therefore, be expected and are found in practice. Nevertheless, they only partially account for the observed degradation.

It was soon found that degradation effects could be reduced by separating the windings, whether this was by using thick insulation or coating with copper or other metallic sheathing (cf. Betterton *et al.*, 1962). It was suggested that degradation was due to some kind of "proximity effect" such as the rearrangement of the mixed state structure (or pinning structure) between H_{c_1} and H_{c_2}, but later work showed that copper was the most effective coating to use to avoid degradation.

The "magnetic history" of the coil when cooled to liquid helium temperatures was also important. Hulm *et al.* (1963) examined a bifilar noninductive coil (300 m of 0.025-cm wire, Nb–25% Zr) and tested it in an externally applied field, with the results shown in Figure 7.5. The applied field was set at fixed values. On their increasing the current in the coil the first superconducting to normal transition occurred at a current I_1, while all subsequent transitions occurred at a value $I_2 \sim \frac{1}{2}I_1$. The curve for I_1 was near that of the short sample. When the current was set first and then the field increased, transitions occurred as indicated by crosses in the figure. Results such as these are consistent with the demonstration (Montgomery, 1962; Aron and Hitchcock, 1962) that after the first self-magnetization there is a residual field left in the solenoid on reducing the applied current to zero. This field reaches a peak near the ends of the coil (Figure 7.6), is in a direction opposite to the main magnetizing field, and is due to trapped flux remaining in the material of the solenoid.

7.2.2. Results up to this point showed that variations in the performance of wire in coils as opposed to short samples must be expected. At the same time the problem of protecting coils when inadvertently quenched had been tackled. One solution was to incorporate shorted copper turns in the coil so as to give the effect of a transformer with shorted turns in the secondary and a long time constant for the decay of fields. The wire could also be copper-plated or, better, copper-clad and such sheathing could be shorted across the ends or at specific points in the coil.

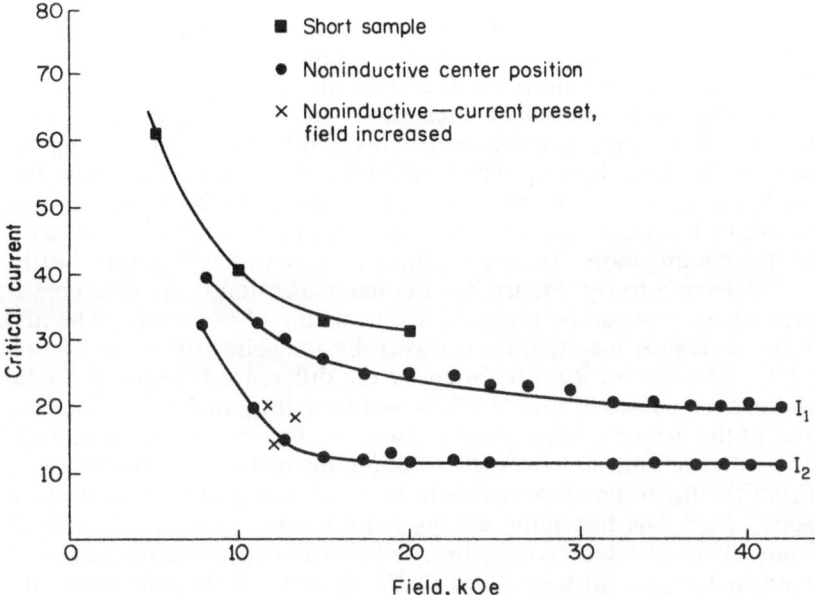

Fig. 7.5. Critical current *vs.* field behavior for a noninductive solenoid.

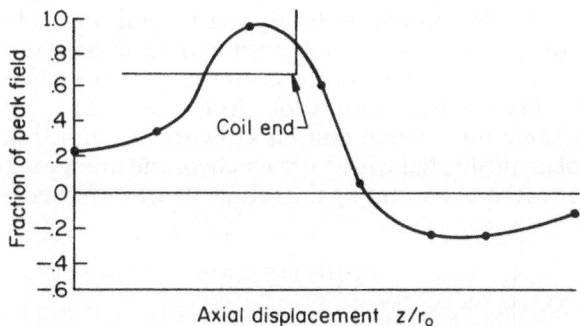

Fig. 7.6. Measured axial residual field profile. Peak field
250 Oe (after Montgomery, 1962).

The most notable advance came with observation (Riemersma, Hulm, and Chandrasekhar, 1964) that the flux in a superconducting coil was not necessarily stable. When a search coil is placed in the bore of a superconducting solenoid and the magnetizing current increased, the signal from the search coil shows the steady increase in field. Superimposed on it however are short, sharp signals with a time duration ranging from 10 to 500 μsec. These are said to be due to "flux jumps"—that is, very rapid flux changes of a few tens of Oersted. Some flux jumps are quite small and do not affect the coil, but others seem to be able to initiate the nucleation of a normal region in the wire. It can then be supposed that normality can be propagated along the wire by the local heating which will take place there. Just how this can happen must be a function of the local geometry, how much copper sheathing is present, and how easily local heat can be dissipated into the helium cooling bath. These are points to which we will return shortly.

Referring to the Figure 7.7, the magnetization curve of a type II superconductor can be regarded as consisting of two parts. The first is the reversible magnetization shown by the solid curve (see Section 6.1.4). The second is represented by the difference between the solid and dashed curves in Figure 7.7, is nonreversible, and is the hysteretic part of the magnetization curve from which flux jumps can be thought of as originating, i.e., from flux-unpinning in bundles (Section 6.2). From the figure flux jumps should be much larger at fields somewhat above H_{c_1}. The flux jump energy is also likely to be larger at such fields. Thus in a superconducting coil we can expect the nucleation of a normal region to take place in those parts of the coil where the magnetic field is low, but nevertheless above H_{c_1}. This idea was investigated by Riemersma, Hulm, and Chandrasekhar, who constructed a coil consisting of 15 concentric cylindrical sections wound from insulated niobium–zirconium wire. By monitoring the individual sections, each time the coil was driven normal the section(s) involved were known. As the field attained at the center increased, so the sections in which normality occurred moved out from the center.

At the same time, Hulm and his co-workers studied the decrease in current of an unshorted coil as it quenched, and attempted a theoretical fit to the curve of current against time. It was necessary to assume

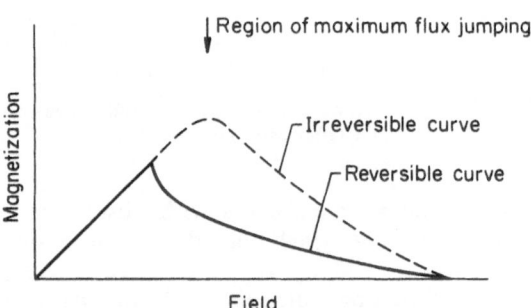

Fig. 7.7. Initial magnetization of a type II super-conductor and flux-jumping.

a small positive initial injection of resistance in the coil for the theoretical results to fit the experimental. Their study also showed that the power derived from the existing battery during quenching is negligibly small compared with that derived from the magnetic energy.

7.3. STABILIZATION OF COILS

7.3.0. Flux jumping has been observed in numerous experiments (Lubell, Chandrasekhar, and Mallick, 1963) and is now accepted as the "triggering" mechanism which can lead to premature quenching of superconductivity in a coil. It is now necessary to consider the advantageous effect of sheathing the superconducting wire with a highly conductive metal such as copper which is usually chosen. Experience has shown that the electrical conductivity of the sheath is important; thus electroplating improves matters but is inferior to a homogeneous sheath. At helium temperatures pure copper has an electrical conductivity which is much higher than that of a hard-worked superconductor in the normal state. When a flux jump occurs, the first effect of the local sheathing will be to increase the time constants for all flux motion over the surface of the superconductor by eddy current damping.

Suppose now that during a flux jump a small section of the superconductor becomes resistive; this may be due to the motion of flux lines without an appreciable rise in temperature, or it may be due to the injection of a normal region due to the dissipation of enough energy. The copper sheath can now provide a low-resistance path for the magnetizing current and, provided adequate cooling exists, any heat dissipated in the copper will be conducted away. In the event that the sheath is too thin, has too low a conductivity, or is inadequately cooled, the heat generated locally will be sufficient to drive the temperature above the superconducting transition and quenching must result. Without a sheath the coil must be prone to quenching, for all flux jumps are then likely "triggers."

In constructing superconducting solenoids the superconducting wire must be used as efficiently as possible, and as high a superconducting filling factor as possible is required. On the other hand, copper sheathing and adequate cooling is required to avoid premature quenching, but just how much is discussed more fully in the next sections. In any event, the designer must provide for the coil to be quenched. For small coils where the stored magnetic energy is only 1 or 2 kJ, such quantities can be dissipated locally without fear of damaging the windings.

7.3.1. As the size of the coil increases and the stored energy reaches tens of kilojoules or more, the design must be carried to the stage of a fully stabilized system; that is, one in which violent quenching is impossible. In some of the very large coils which have been designed or projected, the stored energy is many megajoules. The cost of such coils is so large that the need for inherent stability predominates. Stekly (1965) and his co-workers have shown how this can be done and, indeed, have constructed very large, stable magnets. The principles involved are illustrated by his first constructions. In these the conductor consisted of copper strip 0.1 cm × 1.25 cm in section with nine longitudinal grooves in which nine strands of 0.025 cm diameter Nb–25% Zr wire were inserted. The niobium–zirconium wire was heat-treated and chosen because it had a very high short-sample current-carrying capacity (about 100 A), but in coils showed a very large degradation effect. Such conductors were then wound into pancake coils, which were stacked on the same axis side by side, but enough space was left between each to permit the free circulation of helium. The important result is illustrated in Figure 7.8. As the current was increased, no resistance was seen in the coil until fields just above 40 kOe. On further increase in current, the resistance in the coil rose quite smoothly as shown by the voltage across the coil. On reduction of the current, the coil resistance fell and the curve of Figure 7.8 was retraced with no hysteresis.

What is happening in this cycle is that at low fields all the current is in the superconductor, but when the current reaches the super-conducting critical value for a given field, any further increase in current

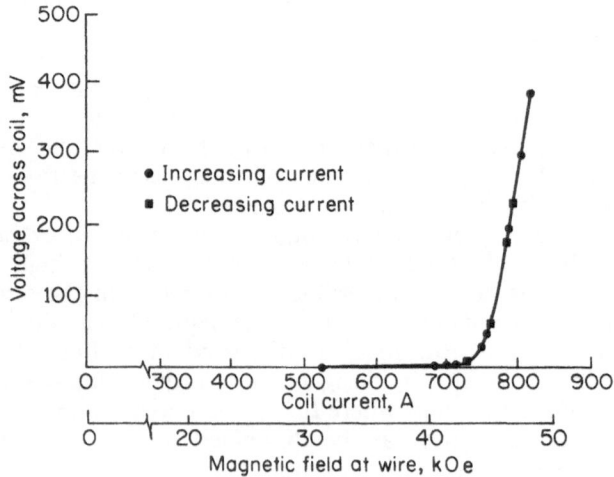

Fig. 7.8. Behavior of a fully stabilized coil (after Stekly and Zar, 1965).

is transferred smoothly to the copper. A small return of normal resistivity in the superconductor must occur so that the potential drops across unit length of the copper and superconductor are maintained equal. As the field increases with the current this transfer must continue until superconductivity is totally destroyed and only a small current remains in the now normal superconductor. If the cooling of the solenoid is sufficiently good, then on reducing the current the same resistance, or voltage–current, characteristic will be shown as with an ascending current.

7.3.2. Stekly and Zar (1965) have given a simplified analysis of the behavior of a system such as has just been described. We imagine a superconducting wire in intimate contact with a normal (copper) substrate or sheath and suppose the fraction of the total current I which flows in the substrate is f. Then $0 \leq f \leq 1$, and the voltage V per unit length of the conductor is

$$V = \rho I f / A \tag{7.6}$$

where ρ and A are the resistivity and cross-sectional area of the substrate. Suppose h is the thermal exchange coefficient, which we will take as constant, between the conductor and the cooling bath at a temperature T_b, and s the cooled perimeter. Then the conductor temperature T is given by equating the heat generated in the conductor to that lost to the bath, and incorporating Equation (7.6),

$$T - T_b = \rho I^2 f / hsA \tag{7.7}$$

The temperature T and the external magnetic field determine the current in the superconductor when current-sharing with the substrate is going on. Now the current-carrying capacity of a superconductor decreases as the temperature rises in a manner which is well approximated by a straight line. Thus, if the maximum current which the superconductor can carry at a bath temperature T_b and in an external field is $I_{ch}(T_b)$, the current I_s in the superconductor can be written as

$$\frac{I_s}{I_{ch}} = 1 - \frac{T - T_b}{T_{ch} - T_b} \tag{7.8}$$

where T_{ch} is the superconducting critical temperature in a field H. To simplify the expressions write the total current I in the composite conductor as a fraction of I_{ch}, i.e., let $x = I/I_{ch}$. Then manipulating Equations (7.6), (7.7), and (7.8),

$$f = \frac{x - 1}{x(1 - \alpha' x)} \tag{7.9}$$

$$\frac{VA}{\rho I_{ch}} = \frac{x - 1}{1 - \alpha'x} \tag{7.10}$$

$$\frac{T - T_b}{T_{ch} - T_b} = \frac{\alpha'x(x - 1)}{1 - \alpha'x} \tag{7.11}$$

In these expressions

$$\alpha' = \frac{\rho I_{ch}^2}{hsA(T_{ch} - T_b)} \tag{7.12}$$

which Stekly calls the "stability parameter." Equations (7.9), (7.10), and (7.11) are shown in Figures 7.9, 7.10, and 7.11. Figure 7.10 shows the voltage–current characteristics of unit length of the conductor. Two distinct types of operation can be discerned.

When $\alpha' < 1$, no voltage appears until $I = I_{ch}$, then as I increases above this value, the voltage increases also and the V–I characteristic for the conductor is everywhere single-valued. For $\alpha' > 1$, however, there is a more complicated situation: first for $0 < I \leq I_{ch}/\sqrt{\alpha'}$ there is a single-valued operation with all the current in the superconductor ($V = 0$). Then for $I_{ch}/\sqrt{\alpha'} \leq I \leq I_{ch}$ there is a double-valued situation with either all the current in the superconductor or all the current in the substrate. Then finally for $I_{ch} < I$, all the current is in the substrate. To summarize, there is completely stable operation for $\alpha' \leq 1.0$, when $\alpha' > 1.0$ stable operation is limited to current up to $I_{ch}/\sqrt{\alpha'}$.

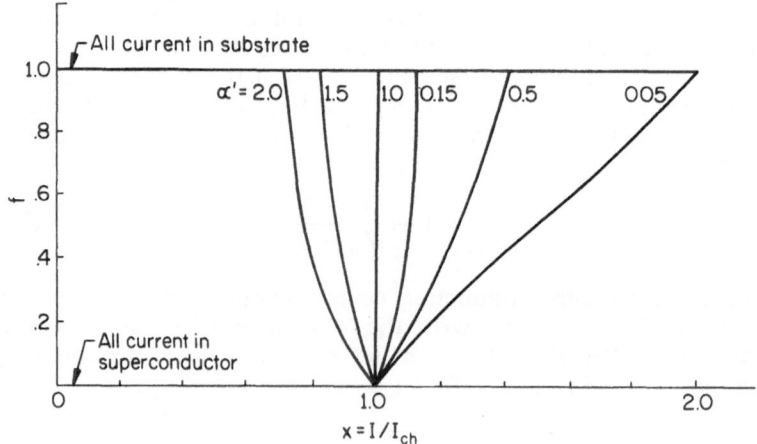

Fig. 7.9. Variation of f with x (after Stekly and Zar, 1965).

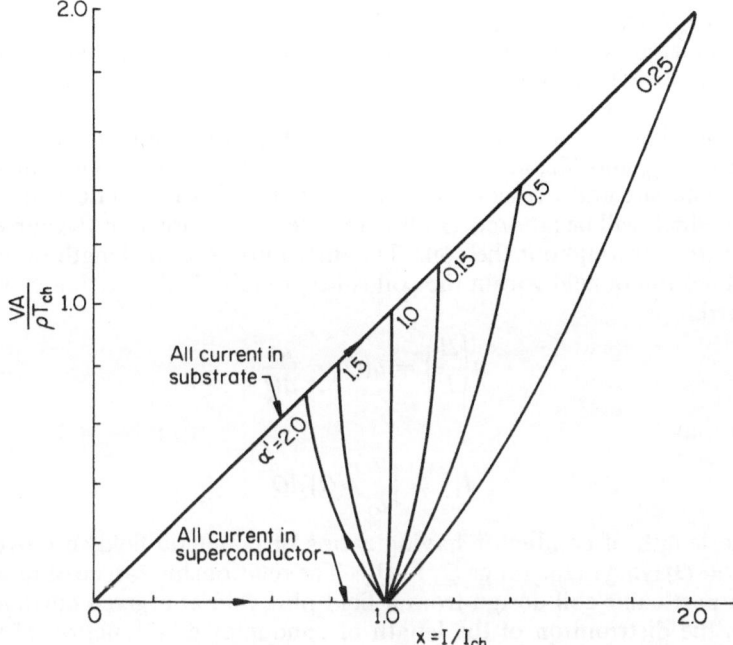

Fig. 7.10. Variation of $VA/\rho I_{ch}$ with x (after Stekly and Zar, 1965).

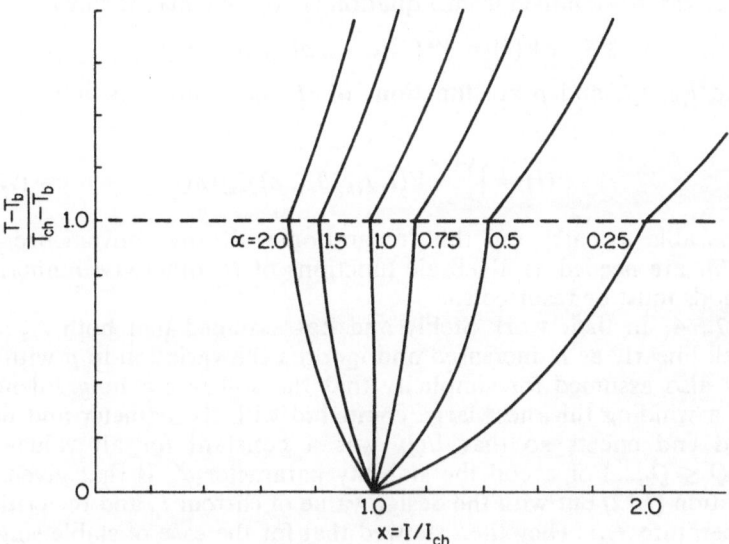

Fig. 7.11. Equation (7.11) plotted for different α.

Above $I_{ch}/\sqrt{\alpha'}$ it is still possible to conceive of steady operation with all the current in the superconductor, but should there be a disturbance, the current will switch to the substrate.

7.3.3. It is now necessary to relate these results to the performance of a conductor in a solenoid where the local magnetic field, and therefore I_{ch} and T_{ch}, are varying from point to point throughout the coil. The substrate resistivity will also vary with magnetic field, an effect which will be ignored. Following Stekly, assume that the current is constant throughout the coil. The distribution of the length of wire as a function of field within the coil is first required. A function can be defined,

$$q\left(\frac{H}{I}\right) = q(Q) = \frac{dl}{dQ}$$

such that

$$l_{12} = \int_{Q_1}^{Q_2} q(Q)\, dQ$$

is the length of conductor having ratios of magnetic field to current $(H/I = Q)$ lying in the range Q_1 to Q_2. The relationship can be obtained for a particular coil design from a field plot, and at a given current it gives the distribution of the length of conductor as a function of the local magnetic field. The increment in voltage dV contributed to the total potential drop across the coil by an element of the conductor in a field H can be obtained from Equation (7.10). We may then write

$$dV(I) = V(I, I_{ch}, T_{ch}, \rho) \cdot q(Q)\, dQ$$

where I_{ch}, T_{ch}, and ρ are functions of H and hence functions of QI and

$$V(I) = \int_{Q=0}^{Q_c} V(I, I_{ch}, T_{ch}, \rho) q(Q)\, dQ \qquad (7.13)$$

To be able to carry out this integration with any convenience, I_{ch} and T_{ch} are needed as algebraic functions of H, otherwise numerical methods must be resorted to.

7.3.4. In their work Stekly and Zar assumed that both I_{ch} and T_{ch} fell linearly as H increased and ignored the variation in ρ with H. They also assumed for simplicity that the coil was a long solenoid with a winding thickness large compared with its diameter and neglected end effects so that l/Q_0 was a constant for all values of $0 \le Q \le Q_0$. For a coil the stability parameter α'_d is that given by Equation (7.12) but with the design value of current I_d and the critical temperature T_{cd}. They then showed that for the case of stable single-valued operation given by $\alpha'_d < 1$ all the current is in the substrate in the inner layers of the coils at high fields. At larger radii the current is shared between the superconductor and the substrate, with the fraction of current in the latter decreasing towards zero with increasing radius. Such a coil should be consistent in its performance (showing no hysteresis) and should operate up to the full short-sample test characteristics of the wire. It should not exhibit uncontrolled transitions to the normal state. For $\alpha'_d > 1$ at low currents, all superconducting behavior will be followed until the current reaches a critical value which is a function of α'_d and the wire characteristics. Above this critical value, double-valued operation will result in which the current is either all in the superconductor or all in the substrate, and when the current is larger than I_d it will all be carried by the substrate. Thus in a solenoid with a high value of α'_d when the current is above a critical value, a disturbance can cause a switch from all superconducting behavior to one in which at least part of the coil is fully normal. To bring the coil back to the all-superconducting state, the current must be reduced until the coil reaches a "single-valued" operating state, and the higher the value of α'_d, the lower will be the value to which the current must be reduced. It is believed that the majority of coils showing large degradation in performance correspond to high values of α'_d. These principles have been applied to the coil which has already been mentioned in 7.3.1, and more recently they have been applied to the construction of a very large coil about 10 ft long (by Avco) designed to give 40 kOe, at which field the stored energy is 5 MJ. Conductor strip of the type described here is designated "Supergenic" strip by the makers.

7.3.5. What this analysis shows is the factors governing the thickness of sheath required and the importance of providing sufficient cooling of the individual conductors. This last point has been further exemplified by the work of Sampson, Strongin, Paskin, and Thompson (1966) on niobium–tin solenoids which showed great degradation effects above 2.18°K, the helium Lambda point, but much improved behavior or behavior corresponding to short sample tests below 2.18°K. At the same time, large flux jumps observable at 4.2°K also disappeared as one would expect, because below the Lambda point the better cooling due to the superfluid helium enables the energy released in a flux jump to be removed very rapidly.

In applying Stekly's analysis the object must be to choose a construction in which α'_d is 1, and the chief difficulty then remains to choose accurate values of h.

7.3.6. Williams (1965) has pointed out that if a geometry is chosen such that α'_d is greater than 1, although the consequent quenching current is less than the short-sample critical value, increased current densities in the solenoid are feasible. He considers the propagation of a normal region in the composite superconductor and defines the "minimum propagating current" I_m as that current at which quenching will just not propagate in the conductor under the given thermal and field conditions. Suppose the cross section of normal material needed to make α'_d of equation (7.12) unity is A_c, then in the chosen conductor

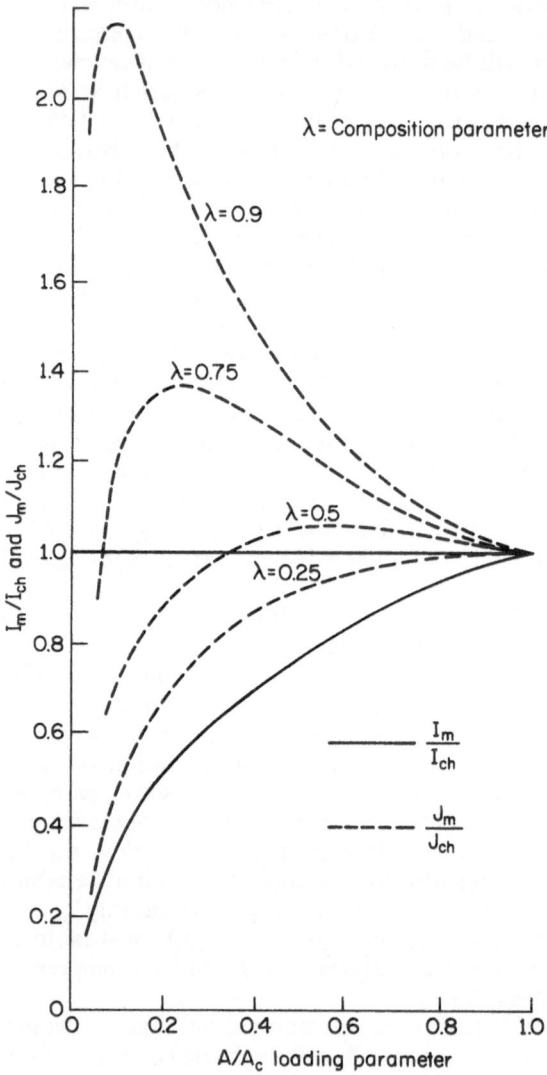

Fig. 7.12. Variation of I_m/I_{ch} with loading parameter (after Williams, 1965).

geometry if the cross-sectional area of normal material is A, the ratio A/A_c can be termed the "loading parameter." The ratio of I_m to I_{ch} (the short-sample critical current) as a function of the loading parameter is shown in Figure 7.12, in which the ratio J_m/J_{ch} is also displayed.

To take into account the cross section A_g of the superconductor itself, Williams uses the composition parameter $\lambda' = A_c/(A_c + A_s)$. Figure 7.12 shows that it is possible for the higher values of λ' to achieve higher overall current densities with a conductor of low loading factor than with a fully stabilized coil. This could be important if minimum weight or volume is the dominant design requirement for a coil. Williams quotes the instance of annealed niobium–zirconium wire with $A_s = 5 \times 10^{-4}\,\text{cm}^2$. Assuming $J_c = 2 \times 10^5\,\text{A/cm}^2$, $h = 0.05\,\text{W/cm}^2\text{°K}$, $(T_{ch} - T_b) = 4\text{°K}$ and $s \sim 0.15\,\text{cm}$, λ' turns out to be 0.93. Thus from Figure 7.12 the maximum current density could be achieved with $A/A_c \sim 0.1$, and the corresponding value of J_m/J_{ch} greater than 2. Provided the highest fields are not being sought, and the energy stored in the coil is relatively small, designs to this kind of criterion could be very economical in material and volume.

7.3.7. The analyses of Stekly and Williams discussed here can be applied equally well to stranded superconducting cables. The superconducting wires may be of niobium–zirconium alloy or of niobium-tin-surfaced wires produced by the methods referred to in Section 6.6.5, copper wires being incorporated in parallel with the superconductors. Cornish (1966), using niobium–zirconium alloys, dipped the composite cables in indium to secure good contact between each member of the cable. Other examples are quoted by Laverick (1965) and Laverick and Lobell (1965).

7.4. COIL CONSTRUCTION

7.4.0. The calculation of coil performance has been outlined in Sections 7.1 and 7.2 which, combined with Chapter 2, enable a coil of a given field contour to be delineated. An additional factor which may have to be included is a reduction of the ratio H_m/H_0 which implies a reduction or elimination of the second order field variation. The current densities and fields in the coil will have been made consistent with the superconducting materials available but the stored magnetic energy will govern the amount of sheathing and much of the constructional features.

7.4.1. If the energy is less than a few kilojoules it is unnecessary to choose a fully stabilized conductor but rather one in which $\alpha'_d > 1$. In this choice, Williams' analysis is a useful guide to the quenching field. For coil formers copper, aluminum, brass, and steel have all been used successfully. As the coil increases in size (outside diameter

> 12 cm) it becomes advisable to match the thermal expansion co-
efficients of the conductor and former. The end cheeks of the former
should be perforated to allow free circulation of helium to the windings
and to enable leads to be taken from the windings which are usually
sectionalized either as concentric cylinders or as pancakes. If the
conductors are not insulated, some form of insulation between layers
is necessary to keep the time constants down. Anodized aluminum
foil has been used by Sampson (1965) in a niobium–tin magnet which
reached 116 kOe with a bore of 8 cm. Mylar sheeting and similar
materials can also be used for this purpose. As the stored energy increases from near a kilojoule more and more precautions must be
taken against quenching. Layers of pure copper foil (e.g., 0.005 cm
thick) to act as shorted turns should also be incorporated at about
every 20 layers and more frequently as the stored energy increases.
Shorting strips of pure copper across all the turns in a layer at frequent
intervals around the circumference are also advisable, their object
being to minimize the chances of arcing due to the high voltages which
would otherwise be induced on quenching. To prevent movement
of the conductors under the action of the magnetic field, strips of tap
grease or rubber compound which freeze hard on cooling, across each
layer, are useful.

   7.4.2. With larger coils still, in the many tens of kilojoules range,
fully stabilized conductors become a necessity. Now the chief design
problem is to arrange for a suitable method of discharging the stored
energy outside the coil should it be driven normal. The simplest
arrangement is to switch an external resistor across the coil, such that
any voltage rise is limited to a safe value. It is usually necessary to
sectionalize a large coil for this reason alone. As an alternative inductive coupling to an external circuit could be used to remove the
magnetic energy.

   Much further useful and detailed constructional information which
is peculiar to their products can be obtained from the manufacturers
of superconducting wires and strip. Useful descriptions of magnet
constructions can be found in the literature (e.g., Sampson and Kruger,
1965).

7.5. LEADS AND CONTACTS IN SUPERCONDUCTING COILS

   7.5.0. The current required to energize a superconducting coil
may be as high as several hundred amperes and it becomes necessary
to optimize the leads which will have to pass through the helium
bath to minimize the loss of liquid helium. The conflicting requirements of low electrical resistivity ρ, and low longitudinal thermal

conductivity k, must be resolved in the design, which also must provide for good thermal exchange with the cold exhaust gas escaping from the helium bath. This optimization problem has been considered by several workers, for example, by Deiness (1965).

For a given current, the ratio of the length of the lead to its cross-sectional area will remain constant. It is, however, advantageous to use a conductor with as low a value of $k\rho$ as possible for the given conditions, and pure nickel is suggested as the most convenient material (at room temperature $k\rho$ is 6.3×10^{-6} V^2/°K, and at 30°K it is 0.12×10^{-6} V^2/°K). By making the leads tubular so that the exhaust gas from the cryostat must pass up them, efficient thermal exchange can be achieved. They can then also be used as structural supports for the coil.

7.5.1. In solenoids of reasonable size, contacts have to be made between lengths of superconducting wire and to leads. If the coil is to be run in a truly persistent current mode, superconducting contacts are necessary. However, a time constant for the decay of the field of 10^6 sec can usually be tolerated except perhaps in nuclear resonance experiments. If the ends of superconducting strip or wire which is well copper- or silver-coated are soldered together with an overlap of several centimeters, the resulting resistance can be as little as 10^{-7} Ω. Such values are quite satisfactory for coils of an inductance of about 1 H. The application of a moderate pressure during soldering to obtain a good bond is frequently recommended, as is the avoidance of overheating.

As coils become larger, external power sources tend to be used and such contact resistances are trivial. With the composite conductor developed by Stekly it has been shown that on winding, many breaks in the niobium–zirconium wires can occur but the lateral resistance through the copper is very small indeed; it is, therefore, not surprising that soldered lap contacts can be used (some 40 cm long).

Simple pressure contacts have also been used successfully (Hulm et al., 1963; Sampson, 1965). The ends of the conductors are brought well outside the coil, where there is a low field, and are clamped between copper or brass plates. Where the plating is left on the conductor, resistances are about 10^{-7} Ω with one or two centimeters overlap; where the wires have been scraped clean, apparently superconducting contacts, at least for currents of tens of amperes, have been made.

7.6. POWER SUPPLIES AND FLUX PUMPING

7.6.0. The currents required in superconducting solenoids range from a few tens of amperes upward to possibly more than 1000 A with the more recently developed multistrand conductors. Power

supplies to give currents require a number of characteristics according to the use of the solenoid. Thus it is frequently desirable to sweep
the current steadily over a wide range and to stop at a steady value.
A typical power supply for a small magnet has been described by
Hulbert and Wilson (1965). For larger coils, particularly where $\alpha'_d > 1$,
a sensing circuit is required to detect the first return of resistance,
so that the power supply can be protected and taken out of the circuit
and a "dumping" resistance switched across the coil. The value of this
resistance must be chosen to prevent objectionably high voltages
appearing (see also Section 7.4.1). Most manufacturers of superconducting solenoids offer suitable power sources to go with them.

7.6.1. Even when optimized leads are used, it is of considerable
advantage to restrict the current to relatively small values (a few
tens of amperes). Attention has therefore been given to methods of
"flux pumping" in which the flux in the solenoid is built up in small
steps. This can be done by using superconducting switching circuits
in a variety of ways. Small prototype systems have been demonstrated
successfully and in consequence solenoid designs in which the magnetizing supercurrent can rise to several thousand amperes are feasible.

7.6.2. Olsen (1958) proposed the use of the all-superconducting
ac rectifying circuit shown in Figure 7.13, in which the current was
introduced by a transformer. The coils PQ and QR wound in opposite
directions and lying in the steady field of another coil C act as rectifiers in a push-pull sense, because the circuit material will become
normal whenever the fields in PQR add to greater than the critical
during a cycle.

In Figure 7.13 the current in the load for a given transformer
excitation is shown as a function of the control-winding current.
Buchhold (1964) described a scheme similar in principle except that
rectification was carried out by the use of "power cryotrons," superconducting switches in which normality can be induced in a niobium

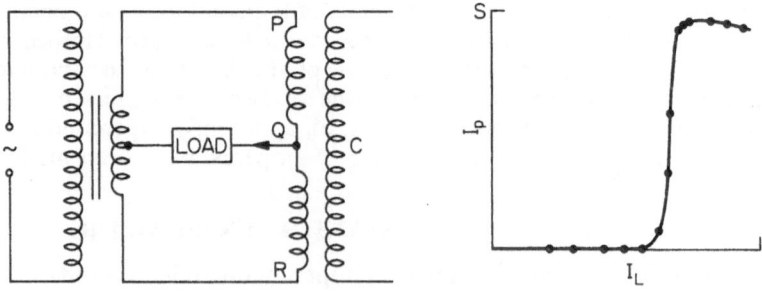

Fig. 7.13. Olsen's superconducting rectifying circuit. I_P = primary current;
I_L = load current.

strip by the magnetic field set up in a secondary winding. The problem with such a scheme is to arrange the timing of the switches correctly to match the waveform of the input power, taking into account the time constants of the cryotrons. Square wave 5 cps was used in the main circuit, which contained saturable reactors from which the signals were used to control the cryotrons. In push–pull circuits similar to Olsen's Buchhold showed that by introducing a phase shift between the operating sequences of the two cryotrons which would normally be in antiphase, any desired level of output current in the solenoid circuit could be achieved. With his prototype circuit the maximum currents achieved were 500 A feeding into a small coil with a stored magnetic energy of about 225 J.

Laquer (1963) devised another, but allied, switching system, the essentials of which are shown in Figure 7.14. S_1 and S_2 are two super-conducting switches which were operated by small auxiliary heaters. With S_2 open and S_1 closed, a flux ϕ was set up in the circuit A by increasing the current in the transformer primary from zero to a fixed value. The switches and transformer were then operated in sequence to share ϕ with the circuit B, set up ϕ in A again, share once more, and continually repeat. A simple analysis shows that the limiting current in the solenoid is that which is induced in the loop A in each cycle. This system has been demonstrated using small solenoids.

7.6.3. A very interesting idea for a "superconducting dynamo" has been developed by Volger and Admiraal (1962), and is further discussed by Volger (1963 and 1965) and by van Suchtelen, Volger, and van Houwelingen (1965). The principle is illustrated in Figure 7.15a. P is a disk of a type I superconductor such that a normal region R can be induced in it by the magnetic field of the small permanent magnet

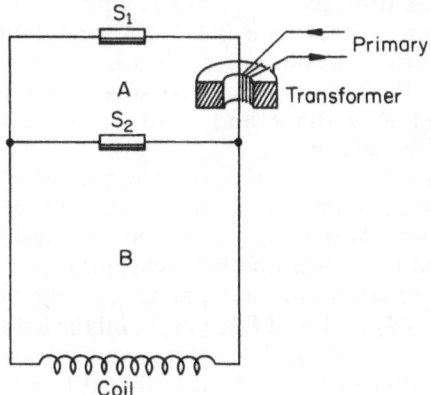

Fig. 7.14. Laquer's switching circuit.

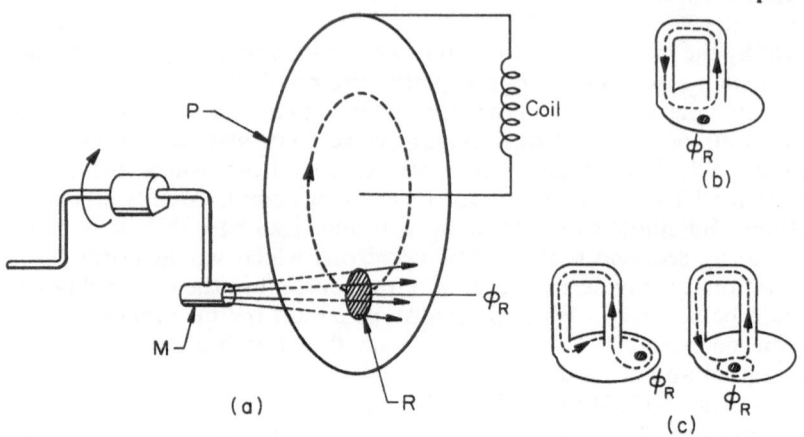

Fig. 7.15. (a) Principle of the superconducting dynamo (after Volger and Admiraal, 1962). (b, c) Integration contours for the flux in a superconducting dynamo.

M, and suppose the flux penetrating R is ϕ_R. M can be rotated across the surface of the disc P as shown, for each revolution the flux ϕ_R passes through the circuit and an induced current appears. After n such passages, the current in the circuit corresponds to a flux $n\phi_R$. That this is so can be seen by considering the conservation of flux in a fully superconducting ring which is part of a multiply connected superconducting body. The topology of the superconductor is defined by the surface of the superconducting phase which excludes the normal region R. Before R is set in motion an integration contour can be taken round the circuit, as shown in Figure 7.15b enclosing no flux. Moving R to pass through the circuit is equivalent to distorting the body and after one passage the integration contour is as shown in Figure 7.15c, enclosing the flux ϕ_R in one loop in the disk and also now enclosing $-\phi_R$ in the circuit. After n passages, the integration contour will loop around R n times and must correspondingly enclose a flux of $-n\phi_R$ in the circuit.

The process is reversible if carried out slowly. Thus if the flux is "wound up" in the circuit, it should be possible for it to "unwind" and drive the magnet M in reverse so acting as a motor. This behavior has been observed by Volger and his colleagues.

If the speed of revolution is f per second, the emf produced can be written as $V = f\phi_Q$. Thus if R is 1 cm² and the field from the magnet M is 5 kOe, V is 10^{-3} V. Such a voltage applied steadily for a time t, to a solenoid of inductance L leads to a current $I = Vt/L$. In designing a system, the time to energize the coil can be specified, so giving the product LI. The stored magnetic energy in the coil, $\frac{1}{2}LI^2$, can also be specified, whence the value of the current required in the coil follows. If the stored energy is taken as 10^3 J, and $t = 10^3$ sec, the current will be 2000 A. The inductance required in the coil can then be calculated. Such systems have been tested with several coils—for example, a solenoid of inductance 10^{-5} H, constructed from Kunzler-type conductors.

For larger coils such as have already been discussed, the emf from the dynamo will have to be increased. This can be done by increasing the size of the flux spot R or the field through it; the former involves a larger machine, and there are limits on the field which can be used. Another possibility would be to connect several elementary dynamos in series, as has been demonstrated by Volger and van Suchtelen, who used 48 coupled together to drive an NbZr solenoid designed for 40 kOe (dynamo emf 0.05 V at 1200 rpm). If the speed of rotation is increased, the emf generated falls below the expected linear dependence on f, an effect which is common to all superconductors used for the disk. The flux spot pattern can be observed visually with aid of a cerium phosphate glass mirror mounted close to it, by using polarized light and relying on the Faraday effect. As the speed increases, a tail develops from the flux spot and "fragments" of flux get left behind largely because of eddy current damping and dissipation due to thermal effects at the spot boundaries. There will also be more complicated effects due to frozen-in flux and inhomogeneities in the disk material Although this technique is still in its early stages of development, the basic ideas are well established and are of such promise that it may well supersede the other methods which have been discussed here.

7.6.4. H. van Beelen and several colleagues (1965) have also published a useful review article on superconducting flux pumps, particularly outlining the systems developed at Leiden. Their flux pump is sketched schematically in Figure 7.16. The circuit consists of a U-shaped thin superconducting sheet S, here represented by a flat sheet, connected across the terminals of a superconducting coil. The pole area of a small U-framed electromagnet is now moved from P to Q with the field on. A region of normal material is created in S and moves with the magnet across it. No net current is generated in the coil circuit by this process. If now the field to the electromagnet is switched off, or it can be brought out of the circuit without disturbing the superconductivity, a current will be generated to conserve flux and the magnet can be moved back to P to repeat the cycle. There are several versions of pumps based on this principle, which have been used to excite fields in the region of 30 kOe, in times as little as a few tens of seconds.

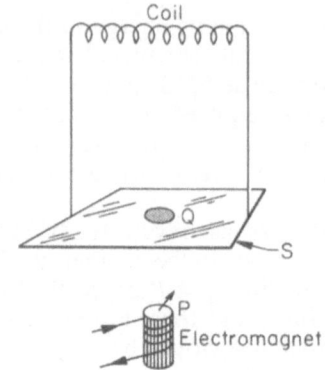

Fig. 7.16. Principle of a flux pump (after
van Beelen *et al.*, 1965).

7.7. FUTURE DEVELOPMENTS

7.7.0. Recently Bean, Fleischer, Swartz, and Hart (1966) have
reported a promising way in which the critical current of a hard
superconductor may be increased. They have added small quantities
(~ 0.3 at.%) of natural uranium and boron to V_3Si and Nb_3Al. The
samples were crushed and sized and irradiated with 1.1×10^{17} and
1.7×10^{18} thermal neutrons/cm^2 to induce fissions or (n, α) reactions.
The fission products produce heavy damage to the crystal structure,
and atoms of intermediate atomic number will also remain in the
material. The uranium-doped samples showed remarkably enhanced
critical currents at 30 kOe, by two orders of magnitude. The critical
current densities of 1 to 2×10^6 A/cm^2 at 30 kOe were approaching
an order of magnitude greater than any hitherto reported for any
material. It must however be clearly understood that these results
were obtained on powdered specimens. Ways must now be found to
repeat them consistently on conductors which can be wound into a
solenoid. If this can be accomplished, it will be a major advance in the
technology of superconductors applied to high-field generation.

Chapter 8

Pulse Techniques and Flux Concentration

8.0. INTRODUCTION

The basic principle of pulsed magnets is that energy is collected and stored in some suitable form, released in short high intensity pulses, and delivered as electrical current through suitable switches to a magnet coil. The most economical form of energy storage depends both on the magnitude and length of the pulse required and on the repetition rate.

Mechanical energy storage in flywheels on rotating machines is most suitable when the length of pulse is relatively long—tens of milliseconds and upward. The usable stored energy can be several megajoules. Lead–acid storage cells can also be used for pulses of similar length and can store many megajoules of energy. New batteries straight from manufacturers are expensive, but second-hand ex-naval submarine cells can usually be bought at near scrap prices and can be used for this kind of duty. Inductors can be used for pulse lengths of a few tens of milliseconds, but usually such systems present switching problems.

The most popular and versatile systems are based on condenser banks, in which energies of several megajoules can be stored. Pulse lengths ranging from a few microseconds up to many tens of milliseconds can be generated by the choice of suitable circuit time constants; the technique is particularly suitable when fast pulses, say 10 μsec, are required. Systems of this kind are cheap and are quite the most important for ordinary laboratory use.

As fields are pushed beyond 250 kOe, the mechanical forces on the solenoids become large enough to break the windings of the coils. With the higher fields, the duration of the pulse must be shortened so that the coil simply has not time to fly apart during the pulse, because

129

of its mechanical inertia. At around 700 k Oe the pulse length must be shortened to about 10 μsec. Even so, coils used in this way have lives which may be only a few hundred pulses long.

Although the power dissipation during a pulse is very high, the pulse must be so short that the temperature rise in the coil is not objectionably large; for this one relies on the "thermal inertia" of the coil. Nevertheless, sufficient cooling must be arranged to dissipate the mean power due to repeated pulsing so that the coil insulation is adequately protected. At or near room temperature air cooling is normally used.

It is a considerable advantage to cool the coil with liquid nitrogen (see Section 8.3.0), for with a given pulse peak power, a much higher field can be generated.

For some experiments it is very useful to concentrate the flux created in a pulsed system into a smaller volume, thus giving a correspondingly higher field. This can be done with some success, and the techniques will be described. They also form the basis for the explosive method in which the very fast release of chemically stored energy (in an explosive) is used to concentrate an already high field (hundreds of kilo-oersteds) into the megaoersted range. By this means, fields of about 10 MOe have been generated in overall pulse lengths of about 10 μsec. Of course, such experiments are essentially "single shot," the experimental "coils" and specimens being lost each time.

In Figure 8.1 the magnitude of the pulsed field is plotted against the practical maximum pulse length with an indication of the principal techniques which can be used. From the historical point of view, reference must be made here to the pioneer work of Kapitza (1924, 1927), whose papers even now give an excellent insight into the practical problems involved in pulsed field work and relatively simple methods of solving them.

8.1. ELECTROMECHANICAL SYSTEMS

8.1.0. Both ac and dc motor generators can be used for this purpose, but they must be specially built to withstand the extra mechanical stresses which will be imposed.

8.1.1. The classic example of the use of an ac generator was the system developed by Kapitza (1927) and originally used by him in his researches at the Royal Society Mond Laboratory in Cambridge. In this system a single phase ac generator was used. A machine with a high peripheral velocity was chosen, thus enabling a large amount of energy to be drawn from the rotating parts in one pulse length. The magnet circuit was made and broken when the current was zero and only one-half cycle was used for each pulse. In principle, the genera-

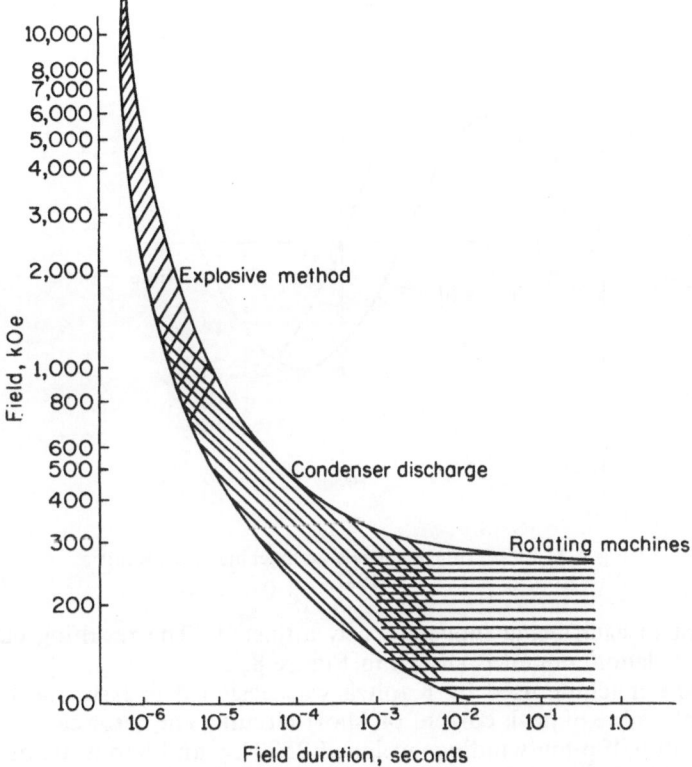

Fig. 8.1. Magnetic field strength as a function of pulse length.

tor could be run up to speed under no-load conditions and the circuit made at a time (shown by A in Figure 8.2) when the emf of the output was zero. During the first half of the pulse, the emf builds up current in the circuit, overcoming the resistance and setting up a field in the coil as well as leakage fields in the machine and circuit. During this time, energy is being taken from the rotor of the machine. In the second half-cycle the emf goes through zero and is reversed, but owing to the magnetic energy stored in the coil and in the machine, the current still persists. This energy is partly dissipated as heat and is partly returned to the rotor of the machine. Kapitza estimated that a fraction as high as 15% could be so returned.

With the simple ac generator system the current pulse is half a distorted sine wave, and it is highly convenient if the top can be clipped off to give a constant current for a short period. For this purpose Kapitza modified the excitation windings. They were divided into concentrated coils one of half pitch and the other of full pitch, the

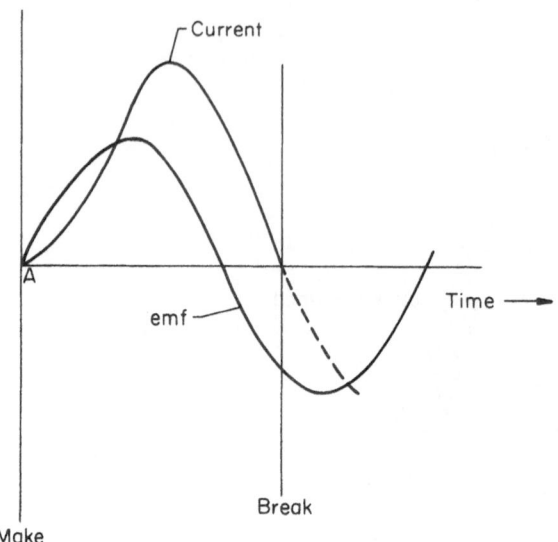

Fig. 8.2. Pulse from Kapitza's ac machine (after Kapitza, 1927).

current in each being independently adjusted. The resulting current in the solenoid circuit is shown in Figure 8.3.

The machine used by Kapitza was designed to have the largest possible value of peak current on short-circuit. The rotor carrying the full- and half-pitch windings weighed 2500 kg, and had a diameter of about 52 cm; its normal running speed was 3500 rpm. With such a moment of inertia, no extra flywheel was necessary. Kapitza (1927) has described his system thoroughly; for the installation details, the reader is referred to his original papers.

One feature, however, should be noted, namely, that switching of the main circuit to the solenoid must be timed accurately. The moment of closing the circuit can be timed by referring to the position of the rotor shaft, for which purpose Kapitza used a cam mechanism. The moment of opening the circuit when the current is zero is, however, variable and depends on the circuit constants. An ingenious mechanical switch mechanism is described in the original paper which enabled the moment of opening to be varied and the correct timing to be found by experiment. A 50-μF condenser was introduced across the circuit at the moment of breaking to avoid the generation of very large voltages due to the collapse of the magnetic fields.

Kapitza successfully generated fields of about 320 kOe in a volume of 2 cm^3 with pulse lengths of about 20 msec, running his machine at 1500 rpm and using only about one-fifth of the power available.

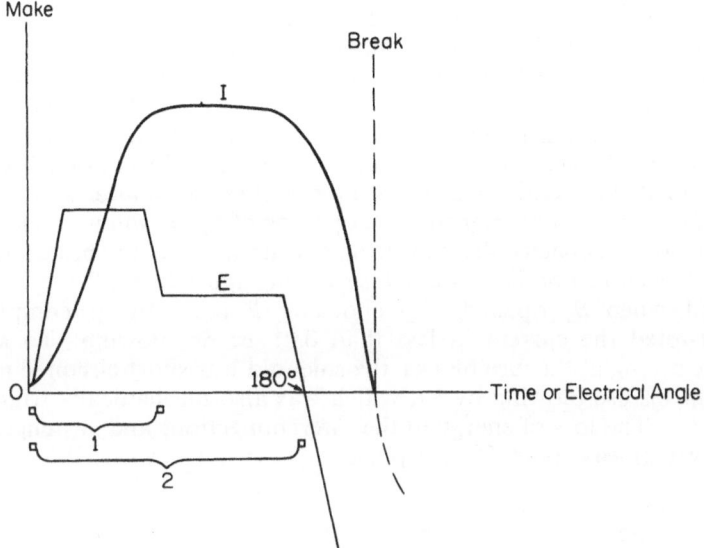

Fig. 8.3. Flat-topped pulse from an ac machine (after Kapitza, 1929).

8.1.2. In the U.S. National Magnet Laboratory the generators (Bitter, 1963) are arranged in pairs, each pair being driven by a synchronous motor and equipped with a flywheel. An auxiliary running-up motor is also provided. The flywheels are 84×10^3 kg in weight and 6 m in diameter with a normal speed of 360 rpm. The continuous rating of the generators is 2 MW, but they are built to withstand pulsed operation for 2 sec, each delivering 8 MW. The energy to be drawn from the flywheels is thus 32 MJ. For pulsed working, the normal design method is to use the running-up motor to bring the generator up to about 385 rpm when the output power level is brought up by overdriving the exciting windings. The machine then slows from 385 to about 340 rpm so that 25% of the kinetic energy is extracted.

8.2. LEAD–ACID CELLS

8.2.0. The first use of lead–acid cells to produce very high fields in a pulsed fashion seems to be due again to Kapitza (1924). For this purpose he designed his own cells with the object of achieving a very low internal resistance, 0.02 Ω for a battery of 70. Using four such batteries connected in pairs in parallel and then each pair in series, a power of 1 MW through an external resistance of 0.02 Ω was realized.

The capacity of the cells was small and the power fell rapidly when the circuit was made. Figure 8.4a shows the variation of power with time. Kapitza constructed his own switchgear for his system, wishing to achieve switching in a time short compared with 0.01 sec, the desired pulse length, for little cost. The system involved three switches. In Figure 8.4b, S_1 was a rapid-action, spring-loaded closure switch which completed the circuit. Using an interlocking mechanical system, the act of closing S_1 also triggered the opening of S_2, a switch of relatively low breaking capacity, at a time from closure which could be accurately set. S_2 was in parallel with a fuse F, through which all the current passed when S_2 opened. By choosing F correctly, it completely interrupted the current in less than 0.01 sec on blowing. To avoid severe arcing at the fuse blocks, the solenoid was short-circuited at the instant of fuse blowing by S_3, which was also mechanically triggered with S_2. The loss of energy in the interconnections and switchgear in this system was about 7% per pulse.

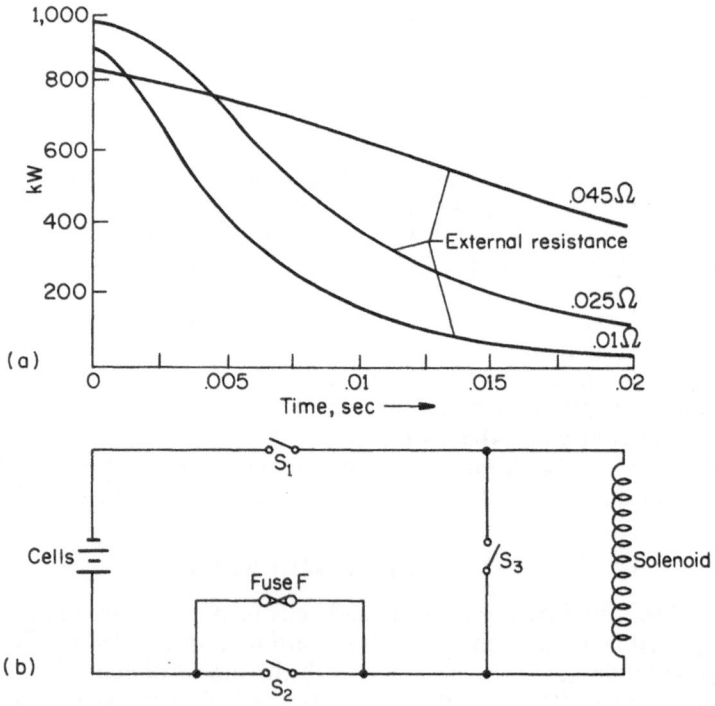

Fig. 8.4. (a) Power as a function of time from Kapitza's lead–acid cells.
(b) Switch system used by Kapitza for lead–acid cells.

8.2.1. It is now possible to obtain from commercial sources lead-acid cells designed for very high discharge duties—for example, single 2-V cells with a standard capacity of about 6500 A-h. 300 such cells fully charged represent an amount of stored energy to be reckoned in kilomegajoules. The internal resistance of a typical cell is about 70 $\mu\Omega$ under normal discharge conditions and it has the advantage that high power can be drawn from it for short periods. Nevertheless, such cells are not ideal for pulsed working, but they can always be so used. In principle, the external resistance of the system should equal the internal resistance of the batteries. Taking into account the initial discharge characteristics of the cells, a system capable of drawing 30 MW for 0.1 sec is feasible. Unfortunately, only about half of this energy will be dissipated in the coil. For a series parallel connection, the working voltage of a system will be 300 and the corresponding currents about 10^5 A. Busbars and circuit components must be strong enough and well enough anchored to support the stresses imposed by such currents. Conventional high-speed breakers can be used to break 10^5 A, but Kapitza's solution is much less expensive, and in any event the coil will have to be short-circuited as the circuit is broken. With powers of 15 MW, fields around 230 kOe are possible in conventional solenoids. With a pulse length of 0.1 sec, the coil can be designed as in the condenser discharge method. If we suppose that a temperature rise of 100°K is permissible during a pulse, a round number calculation shows that the mass of copper which need be used in the coil can be as little as 35 kg. For a reasonable pulse repetition rate, one per minute, water cooling will be necessary, but the conditions are much less stringent than those discussed in Section 3.1 *et seq.* Because of the capital costs, this method is only likely to be embarked on if the need for relatively long pulses is an essential and where a suitable battery bank is available. Skellet (1962) has described a system for energizing coils in the Phoenix mirror machine in which the fields reached are about 120 kOe in coils of 15 cm internal diameter.

8.3. LOW-TEMPERATURE PULSED SOLENOIDS

8.3.0. Where a laboratory possesses good supplies of liquid nitrogen (or hydrogen) but only a small generator or other dc power supply, a "pulsed," cooled coil is a possible and relatively cheap way of generating high fields of long duration (1–2 sec). The idea is to cool the solenoid in liquid nitrogen to reduce the resistivity by a factor 8 or so; then, after removing the nitrogen, to connect the generator across the coil. After the initial large rise, the current will quickly fall again as the resistance rises due to Joule heating. The generator excitation can then be removed, the whole pulse having lasted for perhaps 3–4 sec. The cycle of cooling and reheating with a pulse can then be repeated. The limitation to the performance is set by the heat capacity of the coil.

8.3.1. In such a system, suppose the voltage V applied to the solenoid can be taken as constant, while the inductance of the circuit is L and the resistance is R. R is the sum of the internal resistance of the power supply R_1 and the rapidly changing resistance of the coil which is a function of temperature and will be written as $R_c(T)$. If the temperature of the coil is initially T_0, at a time t after the power has been applied, it will be

$$T = T_0 + \int_0^t \frac{i^2 R_c(T)}{C_c(T)} \cdot dt \tag{8.1}$$

where $C_c(T)$ is the heat capacity of the coil, which also varies with temperature. The current, i, through the coil is given by

$$V = iR_1 + L_1 \frac{di}{dt} + iR_c(t) \tag{8.2}$$

If now $R_c(T)$ and $C_c(t)$ can be given suitable functional forms over the likely limited temperature range of operation, it is possible to calculate the current through the coil as a function of time from Equation (8.2), whence the magnetic field behavior can be found. An analysis of this kind has been carried out by Zijlstra (1962), who used a copper coil working from 77°K. He chose the functions $R_c(T) = 0.032R_{77}(T - 46)$ and $C_c(t) = 0.32C_{77}(t - 46)^{1/3}$ which are good approximations above 77°K. His generator was rated at 100 kW dc at 200 V and 500 A and could be overloaded for a few seconds to 1000 A. His coils were constructed from enamelled wire 6×1 mm with a filling factor of 0.81, an inner bore of 5 cm and length of 12 cm. A field of about 65 kOe was reached and maintained within 90% of this figure for about 0.7 sec. The pulse was discontinued after a few seconds by reducing the generator excitation. In such a pulse, the temperature rose to about 140°K. It took 15 min to cool back to 77°K using 1.5 liters of nitrogen. The extension of such systems to using liquid hydrogen is an obvious step.

8.4. CONDENSER DISCHARGES

8.4.0. The basic circuit for the operation of these systems is that of Figure 8.5. The condenser C is charged from a power supply until the voltage reaches V_0 where upon switch S_1 is closed, thus discharging the condenser through the circuit of resistance R_1 and inductance L_1. During the discharge, the current in L_1 is given by the solution of the very well known equation

$$L_1 C \frac{d^2i}{dt^2} + R_1 C \frac{di}{dt} + i = 0 \tag{8.3}$$

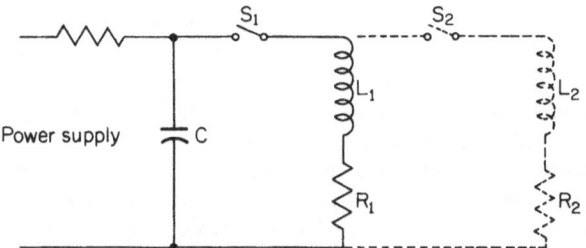

Fig. 8.5. Basic condenser discharge circuit.

There are two ways of operating. First, the circuit constants can be chosen so that the discharge of C is critically damped (i.e., $R_1^2 = 4L_1/C$) or overdamped. The current dies away from its peak with a time constant $2L_1/R_1$ and there is no negative swing on the condenser plates. The highest fields are produced with critical damping but the longest duration with overdamping.

If, however, $4L_1/C > R_1^2$ the discharge is damped oscillatory and the maximum field can be further increased by increasing the frequency. Strong negative swings of voltage occur on the condenser, a condition which is likely to lead to breakdown. This difficulty can be overcome by introducing a second switch S_2 to shunt the coil as the voltage reaches zero during the first part of the discharge. Up to this point, the current i_0 in the coil will be given by

$$i_0 = V_0 \omega C e^{-\beta t} \sin \omega t \qquad (8.4)$$

where $\omega = \sqrt{(1/L_1 C_1) - (R_1^2/4L_1^2)}$ and $\beta = R_1/2L_1$. The equation $\omega t_1 - \tan^{-1}(\beta/\omega) = \pi/2$ gives t_1, the time to reach maximum current in L_1 and zero volts on C.

Closing the switch S_2 introduces an extra resistance R_2 and inductance L_2 into the circuit. The problem then is to choose values of R_2 and L_2 suitable for the task in hand, either to reduce the negative voltage swing on C, say to 20% of V_0, or to reduce the overall power dissipation in the coil L_1. The solutions to the ensuing circuit equations are somewhat cumbersome.

8.4.1. Systems based on these circuits vary considerably in their design according to the duty required of them. For the efficient transfer of energy from the condensers to the coil, the resonant frequency of the capacitor bank must always be considerably greater than discharge frequency ω of Equation (8.4). There is a minimum connection inductance inherent in a capacitor unit for a given container size and an increase in ω can usually be obtained by reducing the capacitance per container. It follows that there will then be a tendency to choose as high an operating voltage as possible, so that the stored energy ($\frac{1}{2}CV_0^2$) is kept high. On the other hand, the higher the voltage the more difficult becomes the insulation problem in the coil itself, particularly where the highest fields are desired. In applications where the coils may be large in volume and the fields relatively low, the insulation problem is usually easier and higher voltages can be chosen safely. A number of systems have been described—for example: Milne, Srivastava, and Wilson (1964), van der Sluijs (1962), Foner and Kolm (1957), Birdsell and Furth (1959), and Shoenberg (1950).

Except for very short time constants the switches S_1 and S_2 are sets of ignitrons; typical circuits for their control and timing are given by Milne *et al.* (above). Petersen (1962) recommends the use of saturable reactors as a means of reducing the total power dissipation in a magnet. This, however, implies a heavy voltage backswing in the condensers and usually means choosing circuit constants to keep ω down to less than 1000 cps. In most solid state applications, the coil sizes required are very small with bores around 1 cm in diameter. A typical low-energy system using 4000 μF charged to 2.5 kV and discharged through copper coils cooled to liquid nitrogen temperatures behaves as indicated in Table 8.I (pulse repetition rate 6 cycles per minute). With more power available and by going to shorter pulse lengths, high fields are obtainable (See Sections 8.4.3 and 8.4.4).

Table 8.I. Typical performance Figures of a 4000-μF, 2.5-kV Condenser

Coil type	Clear bore, mm	Overall pulse length, msec	Maximum field, kOe
Disk	5	0.25	410
Disk	9	0.35	320
Wire-wound	12.5	6.0	250
One-piece helix	25	1.0	150

8.4.2. The small coils used with pulsed systems of relatively low total energy per pulse but designed for high fields are usually cooled by conduction from the outside of the coil. Where possible, it is advantageous to cool the coils to liquid nitrogen temperatures as has been shown in Equation (1.1) and Section 8.3. Direct immersion in liquid nitrogen is to be approached with caution, because liquid nitrogen can permeate into the interstices of the coil. On discharge, the evaporating gas can be trapped and produce sufficiently high pressure to damage a coil. Most coils of the type of Table 8.I can be satisfactorily cooled by gas evaporating from a liquid nitrogen bath. The casing of the magnet coil can be equipped with copper rods dipping below the liquid nitrogen surface in the bath to remove heat by conduction and also to provide gas for gas cooling. The mean coil temperature in a pulse may then be about 150°K, depending on the size of the coil and the energy dissipated in it.

The coils themselves vary in construction according to the duty required. For low fields, coils wound from ordinary enamelled copper wire (e.g., 16 gauge) and "potted" in an epoxy resin are satisfactory. They should be enclosed by a firm-fitting brass case but insulated from it by a fiberglass sleeve. Alternatively, they may be "potted" into a glass fiber case. As the maximum field approaches 180 kOe, such coils become unreliable. Square section conductors are more satisfactory and will work to well above 200 kOe. However, quite the most satisfactory construction is a version of the Bitter solenoid without its cooling holes as shown in Figure 8.6. The outer case which is one electrode must be snug-fitting to the insulating sleeve S and the plates also snug-fitting within S. The end plate of the outer sleeve is drawn down onto the stack so that it is always under strong compression and the end disks cannot pull away from the electrodes. It is unnecessary to incorporate a longitudinal slit in the outer case, which can be made from high-tensile brass, for such eddy currents as may exist have little effect. It is much more important to maintain the mechanical strength of the system. Coils of this type are referred to in the first two entries of Table 8.I.

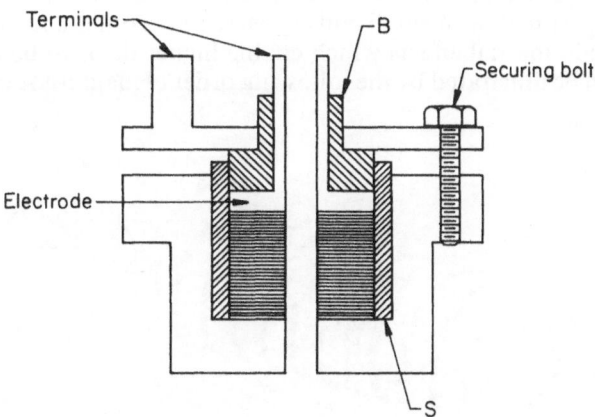

Fig. 8.6. Bitter-type solid disks for pulsed field work.

In this type of construction, care must be taken to see that the insulating washers *B* at the end of the case are always in good condition, for these have to bear the full voltage drop of the system. Under normal conditions, the insulating disks have to withstand only about 50–100 V at the worst. The insulating sleeve *S* may break down when the coil is so overstressed that the individual turns begin to penetrate its inner surface.

8.4.3. Foner and Kolm (1957) have described a rather stronger structure. The coil consists of a flat helix cut from a bar of beryllium–copper. Slotted mica washers were inserted into the coil as insulators in the manner shown in Figure 8.7. The bores of these coils ranged from 2.5 cm down to 4.5 mm. They were clamped firmly between brass end plates and encased in a reinforced ceramic. Coils of 6 mm bore reached fields of 300 kOe successfully and at 4.5 mm bore, fields in excess of 750 kOe. The capacitor bank contained 2000 μF charged to 3 kV, giving an oscillatory discharge with a half period of 120 μsec. For these faster time constants these authors used spark gaps as switches which were comprised of a dual set of adjustably clamped brass electrodes triggered by a third electrode connected to an automobile induction coil. (A 300-V pulse was required on the coil primary winding.) Such electrodes require readjustment after about 10 pulses and refurbishing after about 100 pulses.

8.4.4. Furth, Levine, and Waniek (1957) described a condenser discharge system particularly aimed at achieving the highest fields possible. They used a 24,000-J 4-kV condenser bank discharging into a single-turn coil. They succeeded in generating fields in excess of 10^6 Oe in pulses 10 to 20 μsec long. At fields of this strength all metals must be regarded as fluid or at least plastic, for the tensile limit of even the strongest metals in small coils is passed at around 650 kOe. There then remain inertial effects which enable higher fields to be generated, which can be illustrated by the following order of magnitude calculation

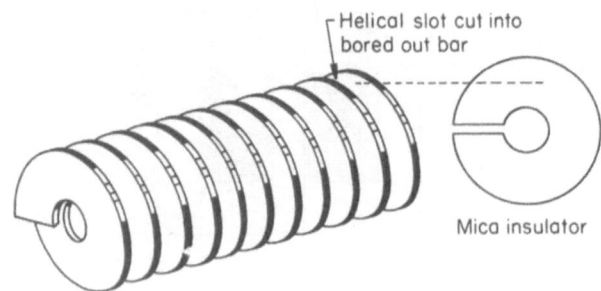

Fig. 8.7. Helical coils (after Foner and Kolm, 1957).

in cgs units. The time, τ, taken to propel a metal sheath through its own thickness t is given by $\tau = t/\bar{v}$, where $\bar{v}$ is the mean velocity. If the field within the sheath is H, the magnetic pressure is $BH/2$, and the mean velocity acquired in travelling out a distance t is $\bar{v} = (BH/4\rho)^{1/2}$ where ρ is the density of the material. If, for example, ρ is taken as $10\,\text{g/cm}^2$ and H as 2×10^6 Oe, $\bar{v}$ is about 10^5 cm/sec. It follows that if a pulse rising to 10^6 Oe lasts for about 10 μsec, the coil (say 1 cm thickness) will stay together just long enough to record an experiment. If in addition it is placed in a very strong and heavy containing sheath, it might even survive.

8.4.5. With fast pulses the skin effect will be pronounced. Once the skin depth δ is small compared with the coil radius, the surface temperature due to Joule heating is independent of the conductivity of the coil or the time scale of the pulse. The resistance of the coil is proportional to ρ/δ, where ρ is the resistivity and $\delta = (2\rho/\omega\mu)^{1/2}$. ω is the "frequency" of the pulse as given by Equation (8.4). The density of heat generated in a pulse is proportional to $\rho/\delta\omega$ and the effective heat capacity is proportional to δ. It then follows that the temperature rise in a pulse is proportional to $\rho/\delta\omega\,\delta = \mu/2$, i.e., is constant and independent of ρ or ω.

Furth and his colleagues point out that with a more precise theory, starting from Maxwell's equations, a diffusion equation can be found for H penetrating into the coil with a diffusivity of $\rho/4\pi$. One can then derive a "characteristic surface temperature" given by $T = H^2/8C_v = 3000\,H^2$, where C_v is the molar specific heat and H is in megaoersteds. These authors also point out that the surface temperature of material of the coil cannot be reduced materially by thermal conduction during the pulse. Good thermal conduction is useful, however, in reducing local "hot spots" due to very small irregularities on the surface of the coil. From this analysis, the conclusion is that at fields above about 900 kOe a thin surface layer of the coil will melt during a pulse, whatever metal is used as the conductor (900 kOe corresponds to tungsten), and surface damage must be expected quite apart from that due to mechanical forces.

To discharge a large capacitor bank into a small single-turn coil efficiently, the conductors and switchgear need careful design. Two large copper sheets separated by Mylar 0.025 cm thick and pressed tightly together acted as the conductors, the current flowing on the inside surface. Wide, flat spark gaps which could be initiated electrically or mechanically were used as switches so that the discharge was about 2 cm in diameter. The coils themselves were typically about 6 mm bore and 2.5 cm outside diameter, with a radial mica sheet providing insulation. They were mounted in a massive hardened steel case in the manner shown in Figure 8.8. It is sometimes convenient to employ

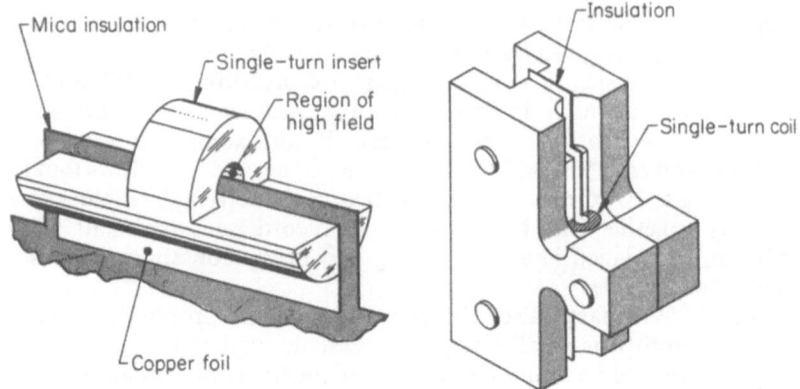

Fig. 8.8. Single-turn coil (after Furth, Levine, and Waniek, 1957).

an air-cored pulse transformer between the capacitor bank and the coil, to increase the current in the latter, and also to insert, by means of a spark gap, a second condenser across the coil at the time of peak field, so that the coil performance is independent of limitations due to the transformer being unable to support more power dissipation.

In carrying out experiments at these very high fields with pulses of a rise time of about 7 μsec, Furth, Levine, and Waniek saw the first damage to the material of the coil at about 650 kOe. Distortion and slight surface melting then increase steadily and at 900 kOe relatively deep pits ($\frac{1}{2}$ mm) can be observed. At higher fields, surface melting effects smooth out the surface damage. If the outer case of the coil does not yield, then as the field approaches 1.5×10^6 Oe, the material of the coil begins to be forced longitudinally out of the case.

8.5. FLUX CONCENTRATORS

8.5.0. A flux concentrator combines the function of a pulse transformer and a single-turn coil but is difficult to make as efficient. The idea is simple and is shown schematically in Figure 8.9(a). A slug of brass A is shown lying inside a single-layer primary winding B. From the cross section in Figure 8.9(b) it will be seen that A contains a radial slot. When a pulse of current is passed through B in the direction of the thick arrows, the current induced in A will flow in its surface due to skin effects as shown by the dotted arrows. The function of the slot is to enable flux (or surface currents) to pass into the center of the slug A, where a high field will be generated. The inner surface of A can be shaped to make the central field a maximum or to give a field of a particular contour.

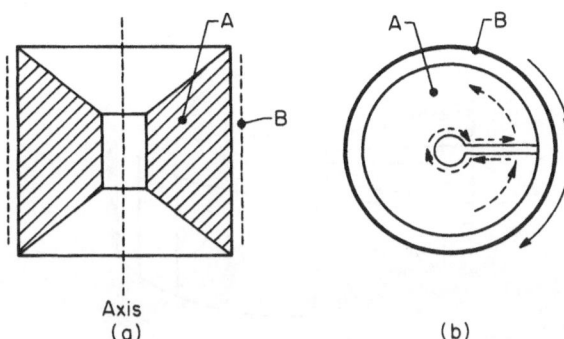

Fig. 8.9. Schematic idea of the flux concentrator.

8.5.1. A single-turn coil has a low inductance and is suited for use with short pulses, but troubles arise from stray inductances. The point of the flux concentrator (or for that matter a pulse transformer) is to make longer pulses easier to achieve and to avoid having to switch very heavy currents. If the primary of the concentrator has n turns and the self-inductance of the slug A is L, the effective inductance as seen from the primary is n^2L, and thus, in principle, the single-turn A can be matched by the choice of n to a given condenser to give a chosen pulse length. It appears from the reports of all users that the energy-transfer efficiency is low, usually 15–20%. An approximate theoretical analysis can be carried out (see, for example, Milne, Srivastava, and Wilson, 1964) from which the circuit losses can be discussed. There are two sources of energy loss: first, there are the resistive losses in the current-carrying surfaces which are a function of the primary and secondary resistances and of the frequency. They can be reduced by using a geometry giving a large winding area. The most important loss, however, lies in the stray inductive energies, for magnetic energy will be stored in those volumes within the circuit which are away from the center of A. This situation can be improved by using a geometry in which the primary winding is embedded in a helical slot in the outer surface of A. The simplest shape for the slug A to take is the hollow cylinder, but in practice this is not efficient. The function of the slug A should be to concentrate the currents induced in its surface as they flow down the slot to give a higher surface density on the inner surface. This can be done by shaping A as shown in Figure 8.10. Some current will flow on the conical surfaces· and represents further losses, but a 3 to 1 concentration can be achieved fairly readily. Milne *et al.* give an extensive analysis of these systems and give results for central fields near 300 kOe. Accurate calculations of the behavior of a flux concentrator are formidable.

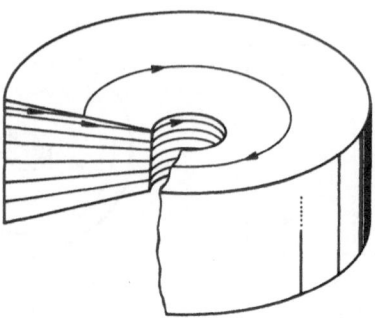

Fig. 8.10. Lines of current in a flux
concentrator.

In Section 8.4.5 it has been shown how with high surface current densities there are acute surface heating effects. These conditions can apply to the inner walls of the concentrator even at fields around 300 kOe. The geometry of a concentrator shows that there are strong repulsive forces in the gap of A and on the primary windings where they cross A, and indeed with most geometries, failure of the primary winding at this point is common.

8.5.1. Brechna, Hill, and Bailey (1965) have described the design, construction, and performance of a large and robust concentrator for use in a high-energy particle experiment (the magnetic moment of Λ hyperons). They required a definite field configuration in which the maximum field at the center of the concentrator was 100 kOe in an area 2 cm in diameter. Using liquid nitrogen cooling, their system survived 4×10^5 pulses at 100 kOe but quickly showed signs of damage at 146 kOe.

8.5.2. Flux concentration can be approached in a different way by considering the interaction of a moving liquid conductor and a magnetic field. If the liquid is a good conductor, it will tend to "drag" the magnetic field lines along with itself. Kolm and Mawardi (1961) used this idea as a basis of a scheme for the continuous compression of flux. In the proposed geometrical arrangement sketched in Figure 8.11, a liquid metal is forced to flow inward radially between the walls of two cylinders in which an axial steady field is being maintained by a set of external coils. The action of the flowing conductor is then to compress the axial field, the amount of compression depending on the velocity of flow and the relative slip between the lines of force and the liquid. These authors have carried out a set of calculations, the details of which are given in their paper, which show the feasibility and difficulties of the scheme quite well. Their results can be summarized in Table 8.II.

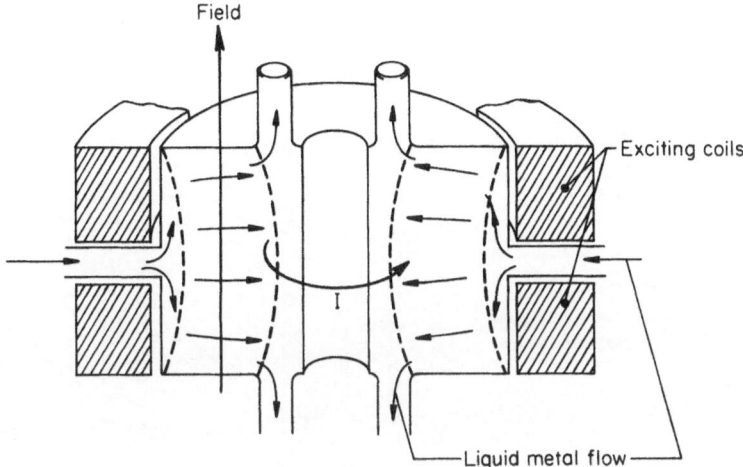

Fig. 8.11. The basis of the "Hydromagnet" (after Kolm and Mawardi, 1961).

Table 8.II

Exciting field, kOe	10	10	60	100	100
Flow rate, liters/sec	37.5	375	22.5	37.5	375
Back pressure, atm	16.5	165	3.6×10^3	1.65×10^3	1.65×10^6
Generated field, kOe	12.4	124	450	124	1240

The row labelled "back pressure" is the pressure which would have to be supplied by the pump driving the liquid metal over and above any hydraulic pressure. The center column of the table represents a set of conditions which might just be reached in the laboratory. There is clearly no cheap and technically easy way of providing the flows and pressures except for limited time periods. In a series of experiments a small quantity (about 2 kg) of liquid sodium–potassium alloy was used as the liquid conductor in an apparatus of which Figure 8.12 shows a cross section. The driving pressure was obtained from 120-atm nitrogen gas cylinders. Applied through the tube A, it forces the liquid alloy on the path shown and radially inward, in the "Hydromagnet" section, which has a mean outer diameter of 6.25 cm and a bore of 2.5 cm. The apparatus was arranged to place this section at the center of a 10-cm bore Bitter solenoid, which provided an exciting field of 64 kOe. The observed enhancement of field was 1.9 kOe compared with a calculated 2.3 for the experimental conditions.

The particular set of experiments, although carried out with elegance and demonstrating the effect sought, confirm that this

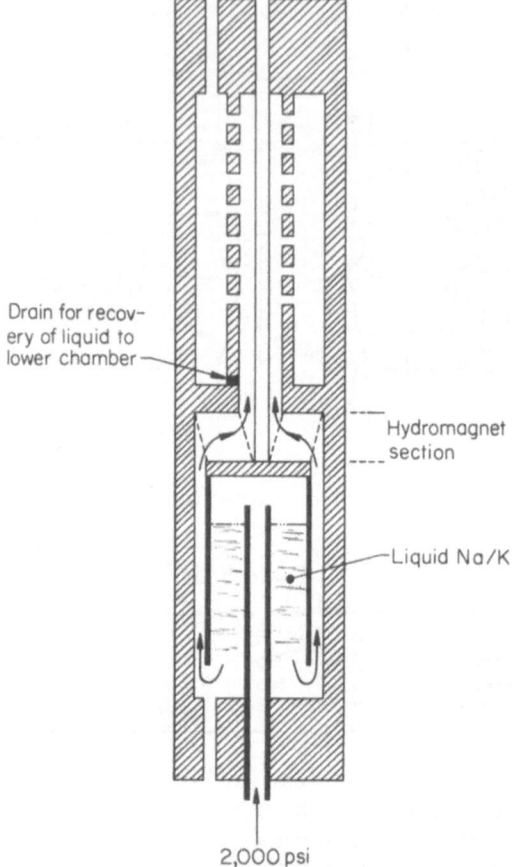

Drain for recov-
ery of liquid to
lower chamber

Hydromagnet
section

Liquid Na/K

2,000 psi

Fig. 8.12. An experimental apparatus for testing the
idea of the "Hydromagnet" (after Kolm and Mawardi,
1961).

method of field compression is unlikely to find a place in steady high-
field technology in the immediate future.

8.6. THE IMPLOSION TECHNIQUE

8.6.0. This technique is another method of flux compression, but
now the forces employed are so great that fields in the 10^7 Oe region
are possible. The principle is simple in concept. In Figure 8.13 imagine
that an initial axial flux is enclosed in a perfectly conducting "liner" A.
If A can now be contracted radially, the flux inside will be conserved,

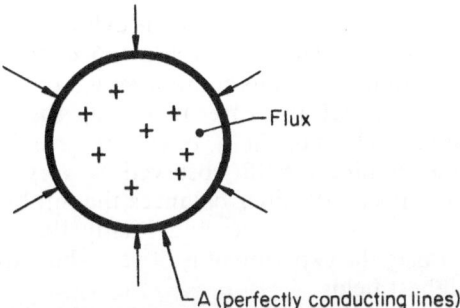

Fig. 8.13. The idea of flux concentration by
implosion.

for it cannot penetrate the walls of A. From Table 8.II we see that if the
fields inside A are going to be large, very large magnetic back-pressures
will exist ranging into the 10^6 atm region. The only way of realizing
this idea in practice is to employ explosives working on the hollow
charge principle and to use a metallic cylinder for A. The resulting
ultra high fields will last for only a very short time, for we have no way
of containing them.

8.6.1. The first report of the successful development of this idea
was given by Fowler, Garn, and Caird (1960), who pioneered its use
into the 10^7 Oe region. The geometry of their system is represented
in Figures 8.14 (a) and (b). A field of 100–200 kOe is produced in a coil
B by the discharge of a condenser bank (150 μF × 20 kV) via a spark
gap. The liner A made of copper or brass has a longitudinal slot milled
into it, so allowing a net flux to be established in A. When the field

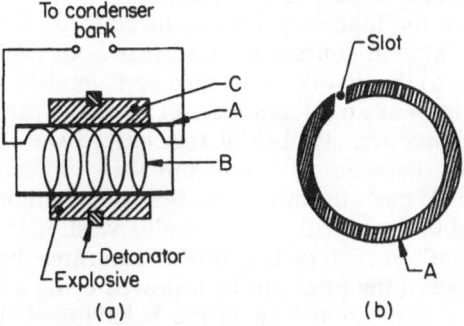

Fig. 8.14. The geometry used in the implosion
technique (after Fowler, Garn, and Caird,
1960).

produced in A has reached a maximum, the cylindrical explosive charge C is detonated so that under ideal circumstances the detonation wave front arrives on A uniformly, thus compressing it radially. The slot in A closes and at the very high pressures involved, A is a continuous conducting fluid liner. The flux in A is compressed into a very small volume, and the resultant field observed is very high. When the magnetic pressure inside the liner balances the implosive pressure, the motion begins to go into reverse and finally the system flies apart violently. Obviously the experiment is of very short duration, of about $2 \mu\text{sec}$ at the highest fields.

8.6.2. With the coils in this construction, the chief point of design is to have them sufficiently well insulated to prevent breakdown to the liner walls. Polythene (~ 0.06 mm) was used for this purpose in the original work. The coils must be as near to the liner walls as possible to minimize the return flux between the coils and liner. The initial working volume within the coils was about 7.5 cm in diameter and 7.5 cm long, while the liners themselves were usually about 3 mm thick. The chief difficulty with them was to prevent sparking across the slot during field buildup when voltages of about 2 kV occurred across it, and to this end it was insulated with scotch tape. As will be seen in Figure 8.14(b), the slot in the liner was milled nearly tangentially. This was found to be the most satisfactory arrangement, giving good electrical closure of the slot when the implosion arrived and also preventing radial metallic jets occurring which could lead to a premature destruction of the experiment.

It is important that the explosive be detonated uniformly, otherwise the liner will be ruptured where the detonation wave first strikes it. The flux then escapes and metallic jets destroy the experiment prematurely. The explosive normally used is "Composition B," which can be machined accurately to fit the liners.

To measure the highest fields produced, probes were introduced axially into the system consisting of a brass body with a central conductor soldered to the tip of a "half-turn coil" as shown. This arrangement could be built as an integral part of a coaxial cable leading to the recording oscilloscopes. Probes of this nature were fitted into glass tubing 8 mm in diameter with a 1-mm wall. Effective probe areas ranged from 0.05 cm^2 upward. Probes were calibrated using the fields generated in the coil B. In an implosive shot, calculation shows that the peak emf's in such pickup probes can range beyond 5 kV.

The collapse of the liner can be followed using a framing camera, which enables a very good idea of the behavior of the system to be built up. The time for collapse to the minimum radius, the turn around radius, is about 10 μsec, as in Figure 8.15. From the probe signals and by calculation from the measured motion, the field–time

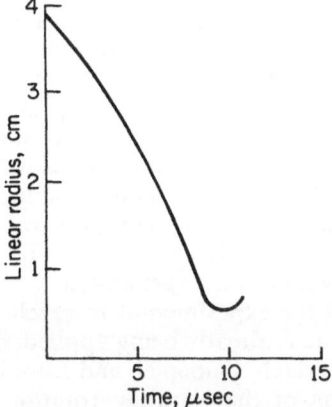

Fig. 8.15. Displacement of liner as a function of time in the implosion technique (after Fowler, Garn, and Caird, 1960).

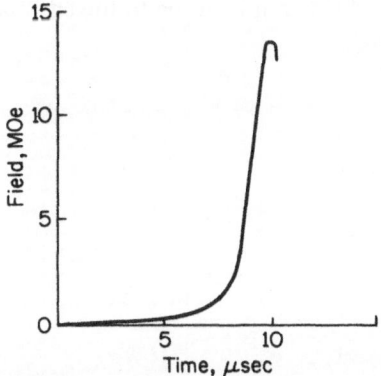

Fig. 8.16. Field *vs.* time curves in the implosion technique (after Fowler, Garn, and Caird, 1960).

curves produced are as in Figure 8.16. Field compression ratios as high as 150 have been observed.

8.6.3. Since this original work other laboratories have also produced fields in the multimegaoersted region, notably the Laboratoria Gas Ionizzati at Frascati (Herlach and Knoepfel, 1965) and also the Kurchatov Nuclear Energy Institute, Moscow (Sakharov, Lyndaev, *et al.*, 1965).

8.6.4. One of the important limitations to the maximum field obtainable by this method could be the finite conductivity of the metallic liner. The difficulty is to estimate the behavior of the conductivity as a function of temperature. At low temperatures it is reasonable to think that it falls off as $1/T$ but to extrapolate in this way to very high temperatures gives field values which are far too low. It is probable that the conductivity of the "metallic plasma" or "liquid metal" decreases to a constant value at the very high temperatures involved (10^4–$10^5\,^\circ K$). With a limited conductivity magnetic field diffusion through the liner during compression certainly takes place, but present thinking is that this is not the limiting factor in determining the maximum fields which can be generated.

8.6.5. As a tool for experimental research, the implosive multi-megaoersted technique is already being applied, particularly to plasma physics. However Herlach, Knoepfel, and Luppi (1965) have reported the first observations of the Faraday rotation in crown glass and quartz in fields up to 2 M Oe. They were able to see up to six complete rotations. Their results provide a very useful confirmation of field strengths deduced from the inductive probes mentioned above. However, the importance of this experiment is that it demonstrates that optical effects of this kind can be followed relatively easily in this type of experiment.

Chapter 9

Future Prospects

In this survey of techniques aimed at generating very high magnetic fields, a number of trends and difficulties on the way toward further progress can be seen.

9.1. SUPERCONDUCTORS

Superconducting solenoids are now the obvious choice for the research laboratory requiring steady fields up to 90 kOe, for these can be bought commercially. This range is already in process of extension toward 150–170 kOe, and we can expect to be able to purchase small solenoids capable of this performance in the near future. Unless a significant increase in the current-carrying capacity of hard superconductors in high fields occurs, an "economic barrier" can already be seen to begin at around 170 kOe. However, one possible way of making very significant advances in current-carrying capacity has already appeared (Section 7.7.0). If there is success from this or similar work, the next barrier with which we are faced is that set by the paramagnetic limit, which with present known materials seems to be near 220 kOe. To find hard superconductors which will have a reasonable current-carrying capacity at higher fields appears to be difficult and expensive. The leads which we get from present theories suggest the further investigation of transition metal alloys, an activity to which considerable effort has already been devoted. Other than this we might consider more exotic ideas such as those of Little (1964) on organic superconductors, but from the practical point of view there seems to be little chance of these "paying off" in the near future.

Superconducting magnets also find a natural place in technological applications. As an example the "magnetohydrodynamic" generation of electricity has already been quoted. Usually for this type of work, a magnetic field of only intermediate strength, say 60 k Oe, is sufficient, but the volume occupied is measured in meters. Similar requirements exist with bubble chambers for high-energy particle research. The stored energy in magnets of such a size is measured in hundreds of megajoules, but fortunately the need for inherent stability can now be met as the result of Stekly's work. The capital cost of such magnets is, however, likely to be very large although the total quantity of actual superconducting material in the composite conductor is relatively small.

We have already pointed out that small superconducting solenoids also have a natural place with those devices or instruments in which there is a need for both high magnetic fields and low temperatures (e.g., the tunable infrared detector, Brown and Kimmett, 1963).

9.2. ORTHODOX COILS

Turning now to the more orthodox methods of producing high magnetic fields, the high-powered, water-cooled solenoid has indeed been superseded by superconductors below about 100 k Oe for laboratory purposes. However, in the field range, which at present begins at 170 k Oe, and in any event will begin at less than about 220 k Oe for a long time, the high-powered solenoid is the only means at our disposal for producing steady fields. In view of the heavy capital cost of an orthodox system for generating such fields, it is natural to think in terms of a large central laboratory containing the necessary facility to which research parties will come to carry out their programs.

It is notable that at the field levels just beyond 220 k Oe the mechanical strength limitation of conductor materials begins to be felt. A method of postponing this limit has been indicated, namely, to accept current distributions such that mechanical forces always remain less than the prescribed limit. The implied reduction in efficiency, or reduced Fabry factor, is a consequence which we have to accept if higher fields reaching toward 300 k Oe are to be obtained; the coils will get larger and more power will be absorbed. It is of interest to inquire how far we can go along these lines. To this end we have carried out a number of outline calculations primarily to get the power-to-field relation into perspective and to estimate size. The results are summarized in Figure 9.1, where the thick straight line represents Equation (1.2) as a standard of reference. The black dots correspond to the performance of known solenoids, while the open circles correspond to calculations which become more speculative the higher the field becomes, and the shaded area gives an idea of the uncertainty of these results. The upshot is that steady fields in the

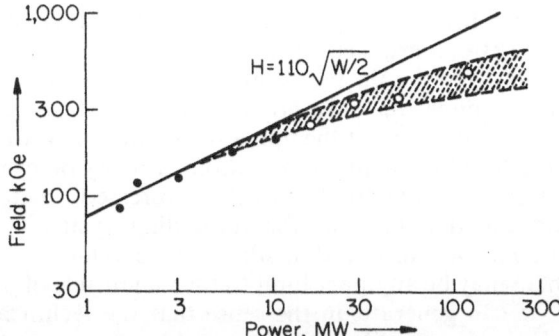

Fig. 9.1. Correlation between power and field in 4-cm-
bore water-cooled solenoids.

400–450 kOe region are technically feasible, but to produce them will
be such an expensive operation from the power point of view that
operation in short bursts (minutes) will be all that can be afforded.

It is relevant to ask whether the use of superconductors can change
this view, and indeed they can. If we can suppose that their high-field
current densities can be so increased that they are an economical
proposition up to say 200 kOe, they should be used for all those parts
of a solenoid which lie in fields less than this. Such use would at least
cut the total power consumption greatly. The inner part of the solenoid
could then be built from a conventional conductor, but designed with a
current density distribution such as to keep always below the tensile
strength limitation. Nevertheless, the total cost of such a "duplex"
solenoid would be great. It is logical to inquire whether the insert for
the superconducting coil could not be cryogenically cooled. Such a
scheme is technically feasible and it would save power for short-period
working, but we now go back to the arguments, largely economic,
already presented in Section 5.3.

Whatever system is used for producing the extreme steady fields,
their duration will be limited by economic reasons of one kind or
another. However, even if this is for only two or three seconds, for many
purposes this is quite long enough to perform steady field experiments.

9.3. PULSED COILS

Beyond about 450 kOe the approach discussed in Section 9.2
above seems to become far too difficult, and one is forced back onto
some kind of short-pulsed-field working, with the duration left as the
only question to be determined. Inevitably this becomes shorter as

the field required becomes higher, and once the peak fields are greater than about 800 kOe, inertial effects dominate the situation as discussed in Section 8.4.4. The implosive compression of flux in a conducting sheath seems to offer the only hope at present of generating fields of many megaoersteds. What the upper field limit is likely to be is open to speculation, for to estimate the situation with any precision requires a knowledge of the conductivity of the compressed "metallic plasma sheath," and the dynamics of the imploding system. The results suggest that 50 MOe will be a difficult target to achieve.

That there may be an upper limit to the magnitude of the magnetic field which we can generate, in the sense that the technical problems involved become insuperable in the foreseeable future, is not surprising, for we are attempting to create one of the few conditions which do not occur naturally in the universe according to our present knowledge.

9.4. FORCE-FREE COILS

At this point, mention should be made of an alternative technique for avoiding the limitations set by the mechanical strength of coil materials. In the analysis of forces in Chapter 2 (Section 2.4) it was assumed that the local flux and current vectors were everywhere perpendicular. If instead they could be made parallel all forces, being proportional to the product $U \times B$, would disappear. To realize this, one seeks appropriate solutions of the equation

$$\text{curl}(H) = KH \tag{9.1}$$

where K is a scalar function of position.

Such an ideal situation has been demonstrated by Furth et al. (1957) to be possible only in an infinite system; it is, however, theoretically possible to design finite systems so that one part of the coil system is force-free while a second, possibly quite separate, part serves to contain the system. A second approach is to design a coil to be force-reduced, rather than entirely force-free. A number of analyses have been published in which methods of solving Equation (9.1) and particular special solutions occur; we should mention particularly those of Furth and his colleagues, Buck (1965), and Wakefield (1964). The latter author particularly is concerned with the actual design of a field-generating system subject to realistic constraints. The same research team has gone further and produced experimental force-reduced toroidal coils (Wells and Mills, 1962).

The potential uses of the technique lie in two directions. The first is in the moderate-field, large-volume applications mentioned above. Because of the physical size, the forces are very great, and because one may also demand accurate alignment the supporting structure becomes very complex and costly. Whether force-reduced systems can really compete depends on many factors, for one must offset the structural savings against the fact that the force-reduced coils may require twice as much power (or twice the volume of superconductor) as their conventional equivalents.

A second, more remote, application lies in the generation of extremely powerful fields. Right up to the economic limit for steady fields, say 400 or 500 kOe, it is difficult to envisage conventional reduced Fabry factor coils being less efficient or more cumbersome than a force-free system. Beyond that, and into the megaoersted region much cruder and cheaper methods of field containment have hitherto proved adequate. Nevertheless, in principle at least, one further barrier to progress is removed.

9.5. CONCLUSION

In the earlier history of the generation of powerful fields, the motivation has come from research laboratories requiring them as tools for solving scientific problems. There has been a very large amount of research activity in fields ranging up to 100 kOe, and in most solid state research involving magnetic fields, many key experiments have been carried out below that figure. Fields in the 100–200 kOe range do not represent a sufficiently large advance for basic research as such to provide the sole reason for investigating the means of generating them. However, once one considers fields well in excess of 200 kOe, the likely dividends in terms of new knowledge undergo an increase, some of which has been indicated in the Introduction. It is into this region that we must go with steady fields which can be held long enough for the more taxing and time-consuming explorations to be carried out. However, on the technological side there are dividends to be reaped over the whole range of fields which we have discussed, whether this be from the point of view of instruments requiring magnetic fields or from the point of view of techniques employing magnetic fields such as the magnetic forming of metals.

References

Throughout *H.M.F.* refers to "High Magnetic Fields," the Proceedings of the 1961 Conference published in 1962 jointly by the M.I.T. Press and John Wiley & Sons, Inc.

Abrikosov, A. A., 1957, *Zh. Éksperim. i Teor. Fiz.*, **32**, 1442 (*Soviet Phys.—JETP*, **5**, 1174).
Adams, C. G., 1962, *H.M.F.*, 33.
Alexander, N. B., and Downing, A. C., 1959, "Tables for a Semi-Infinite Circular Current Sheet," *Oak Ridge National Laboratory Report ORNL—2828.*
Anderson, P. W., 1962, *Phys. Rev. Letters*, **9**, 309.
Aron, P. R., and Hitchcock, H. C., 1962, *J. Appl. Phys.*, **33**, 2242.
Autler, S. H., 1960, *Rev. Sci. Instr.*, **31**, 369.
Bardeen, J., 1954, *Phys. Rev.*, **94**, 554.
Bean, C. P., 1964, *Rev. Mod. Phys.*, **36**, 31.
Bean, C. P., Doyle, M. V., and Pincus, A. G., 1962, *Phys. Rev. Letters*, **9**, 93.
Bean, C. P., Fleischer, R. L., Swartz, P. W., and Hart, H. R., 1966, *J. Appl. Phys.*, **37**, 2218.
Beelen, H., Arnold, A. J. P. T., Sypkens, H. A., V. Braam Houckgeest, J. P., de Bruyn Ouboter, R., Beenakker, J. J. M., and Taconis, K. W., 1965, *Physica*, **31**, 413.
Benz, M. G., Martin, D. L., and Bruch, C. A., 1965, *Cryogenics*, **5**, 248.
Berlincourt, T. G., and Hake, R. R., 1962, *Phys. Rev. Letters*, **9**, 293.
Berlincourt, T. G., and Hake, R. R., 1963, *Phys. Rev.*, **131**, 140.
Betterton, J., Kneip, G. D., Easton, D. S., and Scarbrough, J. O., 1962, *Proceedings of Symposium on Superconducting Materials*, A.I.M.E., New York, Feb. 18, 1962.
Bewilogua, L., and Mahn, H. G., 1963, *Cryogenics*, **3**, 232.
Birdsell, D. H., and Furth, H. P., 1959, *Rev. Sci. Instr.*, **30**, 600.
Bitter, F., 1936, *Rev. Sci. Instr.*, **7**, 482.
Bitter, F., 1939, *Rev. Sci. Instr.*, **10**, 373.
Bitter, F., 1962, *H.M.F.*, 8.
Bitter, F., 1963, *Brit. J. Appl. Phys.*, **14**, 759.
Boom, R. W., and Livingston, R. S., 1962, *Proc. I.R.E.*, **50**, 274.
Bott, I. B., 1964, *Proc. I.E.E.E.*, **52**, 330.
Brechna, H., Hill, D. A., and Bailey, B. M., 1965, *Rev. Sci. Instr.*, **36**, 1529.
Brentari, E. G., and Smith, R. V., 1965, *International Advances in Cryogenic Engineering*, Vol. 10, p. 325, Plenum Press, New York.
Brown, G. V., Flax, L., Itean, E. C., and Lawrence, J. C., 1963, "Axial and Radial Magnetic Fields of Thick, Finite-Length Solenoids," *NASA Technical Report R-170.*
Brown, M. A. C. S., and Kimmitt, M. F., 1963, *Brit. Comm. & Electronics*, **80**, 608.
Brown, W. F., and Sweer, J. H., 1945, *Rev. Sci. Instr.*, **16**, 276.
Buchhold, T., 1964, *Cryogenics*, **4**, 212.
Buck, G. J., 1965, *J. Appl. Phys.*, **36**, 2231.

Chandrasekhar, B. S., 1962, *Appl. Phys. Lett.*, **1**, 7.
Cline, H. E., Rose, R. M., and Wulff, J., 1963, *J. Appl. Phys.*, **34**, 1771.
Clogston, A. M., 1962, *Phys. Rev. Letters*, **9**, 266.
Cockcroft, J. D., 1928, *Phil. Trans. Roy. Soc.*, **227**, 317.
Coffey, H. T., Hulm, J. K., Reynolds, W. T., Fox, D. K., and Span, R. E., 1965, *J. Appl. Phys.*, **36**, 128.
Cornish, D. N., 1966, *J. Sci. Instr.*, **43**, 16.
Daniels, J. M., 1950, *Proc. Phys. Soc.*, **B63**, 1028.
Daniels, J. M., 1953, *Brit. J. Appl. Phys.*, **4**, 50.
Davies, E. A., 1960, *Proc. Roy. Soc.*, **A255**, 407.
Debye, P., 1926, *Ann. der. Phys.*, **81**, 1154.
Deiness, S., 1965, *Cryogenics*, **5**, 269.
Deslandres, H., and Pérot, A., 1914, *Comptes Rendus*, **158**, 226; **158**, 658; **159**, 438.
Desorbo, W., and Healy, W. A., 1964, *Cryogenics*, **4**, 257.
Fabry, Ch., 1898, *Eclairage Electrique*, **17**, 133.
Fabry, Ch., 1910, *J. de Physique*, **9**, 129.
Fakan, J. C., 1962, *H.M.F.*, 19.
Foner, S., and Kolm, H. H., 1957, *Rev. Sci. Instr.*, **28**, 799.
Fowler, C. M., Garn, W. B., and Caird, R. S., 1960, *J. Appl. Phys.*, **31**, 588.
Franzen, W., 1962, *Rev. Sci. Instr.*, **33**, 933.
Friedel, J., de Gennes, P. G., and Matricon, J., 1963, *Appl. Phys. Lett.*, **2**, 119.
Furth, H. P., and Waniek, R. W., 1956, *Rev. Sci. Instr.*, **27**, 195.
Furth, H. P., Levine, M. A., and Waniek, R. W., 1957, *Rev. Sci. Instr.*, **28**, 949.
Garret, M. W., 1951, *J. Appl. Phys.*, **22**, 1091.
Garret, M. W., 1962, *H.M.F.*, 2.
Gaume, F., 1962, *H.M.F.*, 3.
Gauster, W. F., 1960, *Trans. Am. I.E.E.*, I, **79**, 822 (*Comm. & Electronics*, **52**, Jan. 1961).
Gauster, W. F., and Parker, C. E., 1962, *H.M.F.*, 1.
Giauque, W. F., 1926, *J. Am. Chem. Soc.*, **49**, 1864, 1870.
Giauque, W. F., and Lyon, D. N., 1960, *Rev. Sci. Instr.*, **31**, 374.
Ginsburg, V. L., and Landau, L. D., 1950, *Zh. Eksperim. i Teor. Fiz.*, **20**, 1064.
Goodman, B. B., 1961, *Phys. Rev. Letters*, **6**, 597.
Gorkov, L. P., 1959, *Zh. Eksperim. i Teor. Fiz.*, **36**, 1918 (*Soviet Phys.—JETP*, **9**, 1364).
Gorkov, L. P., 1960, Proc. VII Int. Conf. on Low Temperature Physics, Toronto.
Grover, F. W., 1946, *Inductance Calculations*, Van Nostrand, Princeton.
de Haas, W. J., and Voogd, J., 1929, *Comm. Leiden*, 199c.
de Haas, W. J., and Voogd, J., 1931, *Comm. Leiden*, 214b.
Heaton, J. W., and Rose-Innes, A. C., 1963, *Appl. Phys. Lett.*, **2**, 196.
Herlach, F., and Knoepfel, H., 1965, *Rev. Sci. Instr.*, **36**, 1088.
Herlach, F., Knoepfel, H., and Luppi, R., 1965, Proc. High Magnetic Field Conf., Frascati.
Hulbert, J. A., and Wilson, G. R., 1965, *J. Sci. Instr.*, **42**, 293.
Hulm, J. K., Chandrasekhar, B. S., and Riemersma, H., 1963, *Advances in Cryogenic Engineering*, Vol. 8, p. 17, Plenum Press, New York.
Huth, F., 1962, *Cryogenics*, **2**, 368.
Ingram, D. J. E., 1955, *Spectroscopy at Radio and Microwave Frequencies*, Butterworths, London.
Kapitza, P., 1924, *Proc. Roy. Soc.*, **A105**, 691.
Kapitza, P., 1927, *Proc. Roy. Soc.*, **A115**, 658.
Keller, H. B., 1953, *Report UCRL—2249* (University of California Radiation Laboratory).
Klaudy, P., 1962, *H.M.F.*, 21.
Klaudy, P., 1963, *Öst. Ing. Zeit.*, **6**, 154.
de Klerk, D., 1962, *H.M.F.*, 47.
Kim, Y. B., Hemstead, C. F., and Strnad, A. R., 1963, *Phys. Rev.*, **131**, 2486.

Kolm, H. H., and Mawardi, O. K., 1961, *J. Appl. Phys.*, **32**, 1296.
Kronauer, R. E., 1962, *H.M.F.*, 11.
Kuhrt, F., 1954, *Siemens Zeitschrift*, **28**, 370.
Kunzler, J. E., 1961, *Rev. Mod. Phys.*, **33**, 501.
Laquer, H. L., 1962, *H.M.F.*, 13.
Laquer, H. L., 1963, *Cryogenics*, **3**, 27.
Laverick, C., 1965, *Cryogenics*, **5**, 152.
Laverick, C., and Lobell, G. M., 1965, *Argonne National Laboratory, Report No. 7002*.
Léon, B., 1964, *Rev. Générale de l'Électricité*, **73**, 632.
Little, W. A., 1964, *Phys. Rev.*, **134**, A1416.
Livingston, J. D., 1963, *Phys. Rev.*, **129**, 1943.
Löchtermann, E., 1963, *Cryogenics*, **3**, 44.
Lubell, M. S., Chandrasekhar, B. S., and Mallick, G. T., 1963, *Appl. Phys. Lett.*, **3**, 79.
Lüthi, B., 1960, *Helvetica Phys. Acta*, **33**, 161.
Lynton, E. A., 1961, *Superconductivity*, Methuen, London.
McAdams, W. H., 1954, *Heat Transmission*, 3rd Edition, McGraw-Hill, New York.
MacDonald, D. K. C., 1956, *Handbuch der Physik*, Vol. XIV, Springer, Berlin.
Maeda, S., 1962, *H.M.F.*, 46.
Martin, D. L., Benz, M. G., Bruch, C. A., and Rosner, C. H., 1963, *Cryogenics*, **3**, 161.
Maxwell, J. C., 1873, "Treatise on Electricity and Magnetism," Vol. II, Art. 717, Oxford.
Mendelssohn, K., and Moore, J. R., 1935, *Nature*, **135**, 826.
Milne, J. D., Srivastava, K. D., and Wilson, M. N., 1964, *Report NIRL/R/78*, Rutherford High Energy Laboratory, Didcot, Berks. See also: Wilson, H. N., and Srivastava, K. D., *Rev. Sci. Instr.*, **36**, 1096.
Milnes, A. G., 1957, *Transductors and Magnetic Amplifiers*, Macmillan, London.
Montgomery, D. B., 1962, *Appl. Phys. Lett.*, **1**, 41.
Montgomery, D. B., 1963, *Rep. Prog. Phys.*, **26**, 69.
Montgomery, D. B., 1966, *I.E.E.E. Transactions on Magnetics*, MAG-2, No. 3, p. 154.
Montgomery, D. B., and Sampson, W., 1965, *Appl. Phys. Lett.*, **6**, 108.
Montgomery, D. B., and Terrell, J., 1961, "Some useful information for the design of air-core solenoids," *National Magnet Laboratory Report AFOSR-1525*.
Olsen, J. L., 1958, *Rev. Sci. Instr.*, **29**, 537.
Parkinson, D. H., 1962, *H.M.F.*, 41.
Pearson, A., 1962, *J. Sci. Instr.*, **39**, 8.
Peterson, V. Z., 1962, *H.M.F.*, 88.
Pippard, A. B., 1950, *Proc. Roy. Soc.*, **A203**, 210; 1952, *Proc. Camb. Phil. Soc.*, **47**, 617; 1953, *Proc. Roy. Soc.*, **A216**, 547.
Post, R. F., and Taylor, C. E., 1959, *Advances in Cryogenic Engineering*, Vol. 5, p. 13, Plenum Press, New York; see also Post, R. F., *H.M.F.*, 9.
Purcell, J. R., and Jacobs, R. B., 1963, *Cryogenics*, **3**, 109; see also Purcell, J. R., *H.M.F.*, 14.
Purcell, J. R., and Payne, E. G., 1963, *Rev. Sci. Instr.*, **34**, 893.
Rezanka, I., 1963, *Czech. J. Phys.*, **B13**, 545.
Riemersma, H., Hulm, J. K., and Chandrasekhar, B. S., 1964, *Advances in Cryogenic Engineering*, Vol. 9, p. 329, Plenum Press, New York.
Saarivirta, M. J., 1963, *Metal Industries*, **103**, 685.
Saint-James, D., and de Gennes, P. G., 1963, *Phys. Letters*, **7**, 306.
Sakharov, A. D., Lyndaev, R. Z., *et al.*, 1965, *Dokl. Akad. Nauk SSSR*, **165**, 65.
Sampson, W. B., 1965, *Rev. Sci. Instr.*, **36**, 565.
Sampson, W. B., and Kruger, P., 1965, *Rev. Sci. Instr.*, **36**, 1081.
Sampson, W. B., Strongin, H., Paskin, A., and Thompson, G. M., 1966, *Appl. Phys. Lett.*, **8**, 191.
Scott, R. B., 1959, *Cryogenic Engineering*, Van Nostrand, Princeton.
Seraphin, D. P., d'Hurle, F. M., and Heller, N. R., 1962, *Appl. Phys. Lett.*, **1**, 93.

Shoenberg, D., 1950, *Nature*, **170**, 569.
Skellet, S., 1962, *H.M.F.*, 31.
Smith, B. H., Atwood, J. R., and Lyon, D. N., 1964, *Rev. Sci. Instr.*, **35**, 340.
van der Sluijs, J. C. H., 1962, *H.M.F.*, 29.
Stekly, Z. J., and Zar, J. L., 1965, Avco-Everett Research Report 210.
van Suchtelen, J., Volger, J., and van Houwelingen, D., 1965, *Cryogenics*, **5**, 256.
Swim, R. T., 1962, November, *Report of N.R.L. Progress.*
Sydoriak, S. G., and Roberts, T. R., 1957, *J. Appl. Phys.*, **28**, 143.
Thomas, E. J., and Bright, C. D., 1966, *Cryogenics*, **6**, 10.
Vetrano, J. B., and Boom, R. W., 1965, *J. Appl. Phys.*, **36**, 1179.
Volger, J., 1963, *Philips Tech. Rev.*, **25**, 16.
Volger, J., 1965, *International Advances in Cryogenic Engineering*, Vol. 10, p. 98, Plenum
    Press, New York.
Volger, J., and Admiraal, P. S., 1962, *Phys. Letters*, **2**, 257.
Wakefield, K. E., 1964, "Design of Force-free Toroidal Magnets," *Princeton Plasma
    Physics Lab. Report MATT-208.*
Wells, D. R., and Mills, R. G., 1962, *H.M.F.*, 5.
Welsby, V. G., 1960, *Theory and Design of Inductance Coils*, Macdonald, London.
Williams, J. E. C., 1965, *Physics Letters*, **19**, 96.
Williams, F. C., and Noble, S. W., 1950, *Proc. I.E.E.*, **97**, II, 445.
Williamson, K. I., 1947, *J. Sci. Instr.*, **24**, 242.
Wipf, S. L., 1963, *Cryogenics*, **3**, 225.
Wood, M. F., 1962a, *H.M.F.*, 42; see also Daniels, 1950.
Wood, M. F., 1962b, *Cryogenics*, **2**, 297.
Yntema, G. B., 1955, *Phys. Rev.*, **98**, 1197.
Zijlstra, H., 1962, *H.M.F.*, 27.

Subject Index

Author Index